The Calculus of Happiness

the
Calculus of
Happiness

How a Mathematical
Approach to Life
Adds Up to

Health,
Wealth,
and
Love

Oscar E. Fernandez

Princeton University Press

Princeton and Oxford

Copyright © 2017 by Princeton University Press
Published by Princeton University Press, 41 William Street, Princeton, New Jersey 08540
In the United Kingdom: Princeton University Press, 6 Oxford Street, Woodstock, Oxfordshire OX20 1TR

press.princeton.edu

Jacket art courtesy of Shutterstock

ISBN 978-0-691-16863-0

Library of Congress Cataloging-in-Publication Data

Names: Fernandez, Oscar E. (Oscar Edward)
Title: The calculus of happiness : how a mathematical approach to life adds up to health, wealth, and love / Oscar E. Fernandez.
Description: Princeton : Princeton University Press, [2017] | Includes bibliographical references and index.
Identifiers: LCCN 2016044257 | ISBN 9780691168630 (hardcover : alk. paper)
Subjects: LCSH: Mathematics–Popular works.
Classification: LCC QA93 .F467 2017 | DDC 650.101/51–dc23 LC record available at https://lccn.loc.gov/2016044257

British Library Cataloging-in-Publication Data is available

This book has been composed in Minion Pro

Printed on acid-free paper. ∞

Typeset by Nova Techset Pvt Ltd, Bangalore, India
Printed in the United States of America

1 3 5 7 9 10 8 6 4 2

To Emilia

One day when you can read this
Sabrás que estaba pensando en ti
And when that day comes
Búscame y dame un beso y un abrazo
Because there's no better gift
Que una hija le puede dar a su papá.

Contents

■■■■■■■■■■■■■■■■

Preface

■■■■■■■■■■■■■■■■

NUMBERS—THEY'RE EVERYWHERE. They describe the prices you pay for things, the calories you consume each day, and even your dating life (How many people have you dated? How long did those relationships last?). What you may not have realized is that many of the numbers we care about—like the answers to the aforementioned questions—are the *outputs* of various *inputs*. (Example: the calories you consume in a day—the output number—depends on the calories of the individual foods eaten—the inputs.) When thought of this way, an intriguing question emerges: can we use mathematics to improve our lives?

This book is dedicated to answering that question for three aspects of life we think about constantly: health, wealth, and love. Fortunately, many of the input-output relationships in that trio are described by *functions*, special types of equations well-understood by mathematicians. I'll guide you through the formulas behind health, wealth, and love in the chapters that follow. We'll see that they arise quite naturally from careful observations about our everyday experiences, and we'll learn how to extract valuable and powerful insights from them. Here's a sampling of the results to give you a sense of what I mean.

- Chapter 1: A research-backed equation that determines how many calories you should eat each day given certain inputs, including age and weight.
- Chapter 2: A research-based diet that can lower your bad cholesterol, raise your good cholesterol, decrease your risk for developing heart disease and diabetes, and help you lose weight.
- Chapter 2: A research-backed equation that estimates how many years of future life you're losing (and then tells you how to get them back).

- Chapter 3: Several math-backed strategies for increasing your monthly take-home pay.
- Chapter 3: An equation that estimates how soon you'll be able to retire.
- Chapter 4: An investment portfolio whose average annual return since 1926 is 8% during recessions and 10% during expansions.
- Chapter 5: An algorithm to build couples from a group of men and women that *guarantees no one will cheat* with another member in the group.
- Chapter 6: A math-backed way to make joint decisions in a relationship that are perceived as fair and transparent by both parties.

To make the book accessible I've built in several features:

- *Calculations relegated to each chapter's appendix.* Math requires calculations. But rather than force you to read through them, I put them in each chapter's appendix. (Calculations are signaled via starred superscripts, like this one.*[1]) There are some exceptions, but in these cases the details of the calculation contain valuable insights. The main body of the book describes the applications and main concepts of the math and each chapter's appendix details the actual calculations. Moreover, you can skip all the appendixes and still enjoy the book!
- *Online customizable equations linked to book content.* Certain equations have a computer icon next to them (like the one you see in the margin), indicating that I've created online customizable calculators based on the equation. You can access these interactive calculators by visiting the book's website:

http://press.princeton.edu/titles/10952.html

- *Various ways of presenting information.* This book is jam-packed with information. To help you absorb it all, I present information in a variety of ways, including graphically, in tables, with equations, and organized into lists. Anything in *italics* represents information I consider especially important, or is a word or phrase that is being defined.
- *Chapter summaries and tips.* I summarize math and nonmath results at the end of each chapter. I also discuss practical tips related to the content discussed.

- *Very short refresher on background math and notation.* Appendix A contains a quick review of several background (math) concepts you'll encounter throughout the book as well as a glossary of common math symbols.
- *Focus on conciseness.* I'd like you to read this book cover to cover. That's why I've tried my hardest to make it as succinct as possible.
- *Annotated bibliography.* Many of the references I cite—citations are in bracketed numbers (e.g., [4])—are free to read (they're marked as such in the bibliography). For some references I include brief comments on the study's limitations.
- *List of math topics covered.* Following this preface you'll find a list of math topics covered in this book along with the chapter (or appendix) they appear in. As you'll see there, I've organized the book so that the mathematics used gets gradually more advanced as you proceed. I did this to help ease you back into math. You'll also notice that *mainly precalculus-level mathematics is used in this book.* The only exception is in Chapter 6, where I use basic calculus concepts to help describe love dynamics (I provide a brief introduction to the mathematics behind those concepts in the chapter's appendix). This means you almost certainly studied nearly *all* the math discussed in this book at some point in your K–12 education.

Despite my best attempts, there may be times when the material (the math) gets challenging. But don't give up. At those times I suggest rereading the paragraph(s) a few times, pausing in between to think carefully about what I presented. I also recommend that you feel free to contact me. "What?! Write the *author*?" Yes. Here's my email: math@surroundedbymath.com. (If enough people take me up on this offer I may not get back to you as quickly as I'd like to.)

One last thing I'll ask of you: *run any major changes to your life inspired by this book by the relevant expert closest to you first* (e.g., your doctor). These people know the details of your particular situation (e.g., your medical history) and can help you assess how the findings presented here might affect you.

You're about to embark on a journey through mathematics you may not have thought about in a while. As you'll soon discover, math is hidden in things you think about *every day*. Taking the time to understand the mathematics will yield huge rewards, including insights that may dramatically improve your quality of life. I hope this book inspires you to adopt a mathematical approach to life, and helps convince you that math is worth learning. Let's get to work.

Oscar E. Fernandez
Wellesley, MA

Math Topics Covered by Chapter

■■■■■■■■■■■■■■■

Here's a list of the math topics covered in each chapter (or appendix).

Math Topic	Chapter (or Appendix)
Linear functions	1, 3–4, 6
Piecewise linear functions	3
Multilinear functions	1–2
Quadratic functions	1, 4, 6
The quadratic formula	A1
Cubic polynomials	2
Polynomials	A1
Rational functions	1
Exponential functions	3
Comparison of exponential and linear growth	A3
Logarithmic functions	3
Standard deviation	4
Probability	5
Dynamical systems	6
Game theory (specifically, the bargaining problem)	6
3D graphing	A1
Proof of why $1/0$ is not defined	A2
Proof of Gale-Shapley algorithm stability	A5
Brief introduction to derivatives (calculus)	A6

PART I

A Healthier You Is Just a Few Equations Away

CHAPTER 1

■■■■■■■■■■■■■■■■

How Many Calories Should You Eat Each Day?

A FEW YEARS AGO, I used to love going to a chain restaurant—let's call it "Tuscan Fields"—and ordering my favorite dish: fettuccine alfredo. The creamy, cheesy sauce smothering warm strips of fettuccine noodles did it for me, a self-confessed carb lover. I loved the dish so much that at one point I was eating at Tuscan Fields twice a month and making the dish at home. One day I spotted the restaurant's nutritional information pamphlet while waiting for a table. I knew there was nothing in there but bad news. I wanted to continue living my delusion. But I couldn't; I had to know. On page 3 I got the news—the fettuccine alfredo had 1,100 calories and 41 grams of saturated fat!

I was shocked. Every doctor I've had has urged me to keep my daily saturated fat intake under 10 grams. They didn't give me a daily calorie limit to go along with that recommendation, but I was pretty sure that 1,100 calories was too much for one meal. That got me thinking: How many calories should I be eating, anyway? What foods should I avoid, and which ones should I eat more of? There are probably thousands of books on diet, nutrition, and exercise out there that address these questions. But read through the research studies mentioned in their bibliographies and you'll quickly notice something: *mathematics is at the heart of it all.* From calculating "best fits" to the data to determining the error bars, math helps health researchers draw conclusions from their data. If you can understand the math, you have a better chance of understanding the research findings (and their limitations).

That's the premise of this part of the book. I want to show you how to translate the research on nutrition and exercise into mathematics and

how doing so yields valuable insight that cannot be obtained otherwise.[1]
In particular, as we'll see in the next chapter, this approach will help
us build a research-based diet that improves cholesterol numbers,
produces weight loss, and even helps extend life span (no joke—see
section 2.3).

Before we get there we'll need some basic background in nutrition
(and mathematics). That's the goal of this chapter. It will culminate
with an understanding of how many calories we should eat every day,
an invaluable piece of information for all of us. And there's no better
place I can think to start at than a mathematician's favorite venue:
a coffee shop.

1.1 THE LINEAR FUNCTIONS HIDDEN IN YOUR DIET

"Room for cream and sugar?" I'm sure you've heard that line before.
I always say yes (almost instinctively). But today I'll slow things down,
because the cream and/or sugar decisions contain the mathematics that
will build the foundation for the rest of this chapter. But first we need
some basic nutrition facts.

Sugar is a carbohydrate. These oxygen, carbon, and hydrogen (*carbo-hydrate*) molecules yield 4 calories of energy per gram once digested[2]—
at least that's what we've been told. This "conspiracy" can be traced back
to Wilber Atwater, a USDA scientist considered by many to be the fa-
ther of modern nutrition research. Atwater conducted extensive experiments on human metabolism throughout the late 1800s that revealed
wide ranges for the energy yields of each macronutrient.[3] In 1896 he
and his colleagues averaged the yields for each macronutrient, rounded
those numbers to whole numbers, and created the *Atwater general factor system*. The result: the calorie counts now found on every nutrition

---

[1] I'll only include the math I consider necessary to extract the understanding and insights we're
after, and put any additional mathematical details—like calculations—in each chapter's appendix.

[2] North Americans call them "calories," but they're actually *kilo*calories, which explains why
food labels in other parts of the world show "kcal."

[3] For example, he found that carbohydrates can yield anywhere between 2.7 and 4.1 calories per
gram depending on the food (e.g., white bread vs. rolled oats).

label in the United States (and other parts of the world)—carbs and proteins yield 4 calories per gram, and fats yield 9 calories per gram [1].

Four calories for each gram of sugar isn't a lot. But it means that the added calories will be *four times* the total sugar I add to my coffee. Specifically, if I add x grams of sugar that will add $4x$ calories.[4] Denoting by y the calories added, we get

$$y = 4x.$$

Voilà—we've just derived our first equation!

Now let's introduce some terminology. Since both x and y can vary, they're both examples of *variables*. To find y we need to know x, so we call y the *dependent variable*. (Indeed, the value of y *depends on* the value of x: for example, if $x = 4$ then $y = 16$, whereas if $x = 3$ then $y = 12$.) Similarly, we call x the *independent variable*, since its value doesn't depend on any other variables in the equation.

But $y = 4x$ only yields useful insights if we recognize it as a linear function, a family of functions that will come up often in this book. I explain what a function is in Appendix A, but since every equation we'll discuss defines a function, there's no need for you to know the precise definition. (For that reason I use function and equation interchangeably.) That $y = 4x$ is a *linear* function, however, is more important for our purposes. Here's the simplest definition of a linear function.[5]

How to Spot a Linear Function

An equation of the form

$$y = mx + b$$

is called a *linear function*. The number b is called the *y-intercept*, and the number m the *slope*.

---

[4]A quick remark about notation: "$4x$" means "four times x."
[5]The most general definition of a linear function is an equation of the form $Ax + By = C$. This allows for cases that result in vertical lines (when $B = 0$). But we won't encounter that in the book.

Comparing our "sugar function" $y = 4x$ to $y = mx + b$, we see that $b = 0$ and $m = 4$. The y-intercept is easy to interpret: if I use no sugar ($x = 0$) then I add no calories ($y = 0$). To interpret the slope, picture me pouring the sugar—very slowly—into my coffee. Each granule of sugar that comes out increases the x-value (the number of grams of sugar added to my coffee). When I get to 1 gram of sugar poured in, I've added $4(1) = 4$ calories. From there, each additional gram I pour in increases the calories by 4, the slope of our sugar function.

Aha! The slope of our sugar function is telling us how much the y-value increases (i.e., how many calories are added) when the x-value increases by one unit (i.e., when 1 gram of sugar is added). This interpretation for the slope holds true for a general linear function, too[*1] (recall that starred superscripts point to calculations or other mathematical details in the chapter's appendix). More formally, here's the generalized slope interpretation: *when the x-value increases by one unit, the y-value of a linear function increases by m (if m > 0) or decreases by m (if m < 0).*

Our sugar function's "right 1, up by 4" line dance can be visualized by graphing the function $y = 4x$. This is a useful way to *see* functions, so let me review how to construct the graph in Figure 1.1(b).

First, we create a table of values for x and y (the first two columns of Figure 1.1(a) show a few examples). Then we draw a set of axes perpendicular to each other—forming the xy-plane—and call their intersection the "origin." We set a scale for the horizontal x-axis and the vertical y-axis (in Figure 1.1(b) the horizontal tick marks are spaced 0.5 unit apart and the vertical tick marks are spaced 5 units apart). Each location in the grid within these axes now has a particular x-value and a particular y-value (e.g., "5 units to the right of the origin and 20 units above the origin"). We combine these two values into the point (x, y), and then plot these points on the grid (the third column of Figure 1.1(a) shows a few examples of points; they're plotted as the dots in Figure 1.1(b)). Finally, we connect the dots to get the graph of the function (Figure 1.1(b)).

Constructing linear functions from given information, interpreting their slopes, and being able to read their graphs are essential skills we'll use frequently in this book. My creamer decision provides

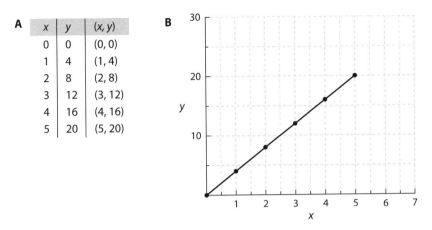

x	y	(x, y)
0	0	(0, 0)
1	4	(1, 4)
2	8	(2, 8)
3	12	(3, 12)
4	16	(4, 16)
5	20	(5, 20)

Figure 1.1. (a) A table of values and points for the linear function $y = 4x$. (b) The graph of $y = 4x$ for x-values between 0 and 5 (the maximum amount of sugar in one packet); the dots are the points in the table.

another example. Adding x tablespoons of cream (say, half-and-half) to my coffee would increase the caloric content by $9x$ calories.[6] The higher slope here means that one tablespoon of creamer adds *more than double* the calories of 1 gram of sugar. No, thanks; I'll skip the cream today.

I've now made my cream and sugar decisions, but let me "mathematize" another scenario (that's my phrasing for the process of translating information into math), since it'll set the foundation for what we do in the next section. This one's motivated by the chocolate croissant I'm staring at: how much sugar can I add to my coffee if I plan on eating that chocolate croissant but want to limit the total calories to 400?

The little label I see next to the croissant says it contains 370 calories. So the total calorie count, which I'll denote by c, is 370 plus the calories from the sugar I add to my coffee ($4x$). That gives the linear function

$$c = 4x + 370.$$

The slope is again $m = 4$, but this time the y-intercept is $b = 370$. (If I add no sugar to the coffee my meal would still contain 370 calories.)

---

[6] I'm assuming that all the creamer's calories come from fat, which is nearly true.

Keeping the total calories under 400 is equivalent to $c \leq 400$ ("c less than or equal to 400"), whose solution is $x \leq 7.5$ grams.*2 So I can safely add a little sugar to my coffee *and* eat the croissant while keeping to my 400-calorie cap.

These analyses illustrate one of the many benefits of "mathematizing" a situation or problem: you get to use all the results and techniques mathematicians have developed to tackle the problem. Often this leads to new insights; in our case we discovered that the energy yields of carbs, protein, and fat are slopes of lines. (That's the reason for this section's title—slopes and linear functions are hidden inside every bite you take!) Sometimes "mathematization" leads to new applications. For example, the same approach to our coffee + croissant problem can help someone estimate how much protein, carbs, and fats they can eat while keeping to a prescribed calorie cap.*3

Now that we have a working understanding of the mathematics of calories, let's move on to the mathematics of metabolism.

1.2 THE MATHEMATICS OF METABOLISM

From the corner where I'm sitting, I see slim people downing huge Frappuccinos (these can top 600 calories). Why is it that some people can eat more calories than others and not gain weight? That question has a complicated answer, but let's see what math can do for us this time.

First, a few observations: the people sipping on those highly caloric drinks are tall and young. One guy sitting next to the coffee station—a few tables from me—is about 6 feet tall. He looks to be in his early 20s, with an athletic build. The barista just greeted him by name—Jason. Jason probably needs more calories to maintain his weight than most of us given that he's tall, young, and athletic. But is any of this true? In short, yes.

Each of us has a *resting metabolic rate*, or *RMR*. By definition, this is the *daily* energy (calories) your body would burn in an awake, nonfasting, at-rest state.[7] Translation: your RMR is the daily energy

---

[7] RMR measurements are made a few hours after a light breakfast [2].

needed by your body to complete its normal tasks (e.g., circulating blood). That should make you happy, because it means that *every day you burn your RMR in calories without having to lift a finger!*

To find out your own RMR you could visit a lab and have a sports scientist hook you up to an *Atwater-Benedict-Roth apparatus* (yup, that's the same Atwater from before); your RMR would then be determined by measuring how much oxygen you consume in a 6-minute period (while at rest). But that visit will cost you. Plus, it's a black-box answer—it doesn't tell you what variables RMR depends on.

That's where we come back to Atwater. His experiments kicked off a century of research into the science of metabolism, and one of the first things to be quantified was RMR (by Harris and Benedict in 1918). They came up with a formula that depends on just three variables: your weight, height, and age. Subsequent experiments led to more accurate equations. A recent comparison of the available formulas crowned the winner: the 1990 *Mifflin–St. Jeor* equations [3].[8]

Estimating Your RMR with the Mifflin–St. Jeor Equations

$$\text{RMR}_{\text{m}} = 4.5w + 15.9h - 5a + 5, \qquad (1.1)$$

$$\text{RMR}_{\text{w}} = 4.5w + 15.9h - 5a - 161. \qquad (1.2)$$

The first equation estimates the resting metabolic rate for men, the second for women, and both assume you're at least 19 years old. In both equations w is weight in pounds, h is height in inches, and a is age in years.

(Remember, I've created online calculators on the book's website for formulas with computer icons next to them.)

The first thing you may have already noticed is that the equations are almost the same; the only difference is the last term ("+5" versus "−161"). In fact, adding 166 to (1.2) gives (1.1). Thus, the

---

[8]These formulas aren't perfect; see the bibliography for a short discussion of the researchers' comments regarding their accuracy.

Mifflin–St. Jeor equations predict that *a man needs 166 more calories than a woman of the same weight, height, and age.* Score another one for new insights from math!

The other numbers in the equations, the "coefficients" 4.5, 15.9, and −5, also have a story to tell. Let's discover their meaning by using Jason as a guinea pig. Let's say he's 23, so that $a = 23$. Since he's 6 feet tall, $h = 72$ inches. Plugging in these values into (1.1) gives

$$RMR_{Jason} = 4.5w + 1,034.8. \qquad (1.3)$$

Aha! That's a linear function (with slope 4.5). Had we gone back to (1.1) and plugged in Jason's weight and age instead, we'd have been left with another linear function, this time with slope 15.9. So, the Mifflin–St. Jeor equations contain *multiple linear* equations as special cases. *Indeed, the RMR equations are examples of multilinear functions.* I would even describe them as "linear in the w-variable with slope 4.5," "linear in the h-variable with slope 15.9," and "linear in the a-variable with slope −5." (The graphs of multilinear functions require three or more dimensions; check out the 3D graph in the appendix[*4] for an illustration.)

But wait, there's more! Now that we've connected the Mifflin–St. Jeor equations with linear functions, we can apply our slope interpretation to the coefficients in (1.1)–(1.2). Focusing on the 4.5 coefficient of w, for example, the equations predict that *for every pound you gain, your RMR goes up by 4.5 calories.* Table 1.1 summarizes the insights gained by interpreting the other slopes.[9]

Table 1.1 suggests that Jason, who's busy chatting with the barista, does indeed have a high energy requirement—he's tall and young, both of which raise his RMR. The older gentleman sitting across from me who looks short and slim, however, likely has a lower RMR. In fact, now that I'm looking around at everyone else in this coffee shop, I can almost "see" each person's RMR, as if each had a speech cloud containing their

---

[9]Since the RMR_m and RMR_w equations have the same coefficients, the conclusions of Table 1.1 apply to both men and women.

TABLE 1.1.
The predicted effect on RMR of a one-unit increase in weight, height, or age. The direction of RMR change is reversed for one-unit decreases (e.g., "up" becomes "down").

If you . . .	Your RMR should go . . .
Gain 1 pound	Up by 4.5 calories
Grow 1 inch taller	Up by 15.9 calories
Age 1 year	Down by 5 calories

particular number. So not only are linear functions hidden in every bite you take, multilinear functions are hidden *inside of you*.

Now remember, RMR is defined relative to an at-rest state. If we so much as get up and go for a walk, we're burning more calories than our RMR predicts. In the next section we'll learn how to quantify those extra calories.

1.3 BURN THOSE CALORIES! WORK THOSE QUADS!

Most of us associate the word *exercise* with a physically demanding activity, like running or swimming. But any activity requiring physical effort constitutes exercise. That means even the busy baristas making drinks in this coffee shop are exercising. They may not be huffin' and puffin', but by the end of the day they'll have burned more calories than their RMR.

To quantify that extra caloric burn, let's focus on *aerobic* exercise (i.e., exercise that requires you to breathe faster than normal). Faster breathing means more oxygen consumption, and thanks to those Atwater-inspired experiments, we know that about 5 calories are burned for each liter of oxygen consumed. (It's no wonder aerobic exercise is the most effective way to get rid of body fat [4].) Sports scientists use this factoid to estimate an individual's *aerobic caloric burn* (ACB)—the calories burned per minute of aerobic exercise. Like RMR, researchers have come up with several equations to estimate ACB.

Here's a relatively accurate one that's a function of weight, age, and heart rate [5]:

Estimating Calories Burned per Minute of Aerobic Exercise

$$ACB = 0.02w + 0.05a + 0.15r - 13. \qquad (1.4)$$

Here ACB is calories burned per minute of aerobic exercise, w is your weight (in pounds), a your age (in years), and r your heart rate (in beats per minute [bpm]).

There are several neat things to notice about this formula. The first is that it applies to all types of aerobic exercise.[10] Another feature is something I hope you've already noticed: it's another multilinear function! That means we can interpret the coefficients of w, a, and r in much the same way we did for the RMR equations. Some of the ensuing insights are obvious (e.g., the higher the heart rate, the higher the ACB), but one result seems counterintuitive: the older you are, the higher the ACB. The multilinear nature of (1.4) also means we can use the same approach that reduced the Mifflin–St. Jeor equations to a linear function to make the formula more useful for practical applications. For example, let's plug in Jason's age ($a = 23$) and weight (about $w = 150$). This reduces (1.4) to

$$ACB = 0.15r - 8.85, \qquad (1.5)$$

a linear function. The 0.15 slope says that for each additional 1 bpm Jason's heart beats, he burns 0.15 more calories per minute. To put this in context, let's say Jason wanted to burn the 400-calorie coffee + croissant snack I ate earlier in 20 minutes. That's an ACB of 20 calories per minute, and according to (1.5) he'd need to sustain a heart rate of about $r = 192$ bpm for those 20 minutes to burn off that snack.[*5] That's very high (a racing pulse is as low as 140 bpm). In fact, 192 bpm is higher than Jason's theoretical "maximum heart rate."

---

[10]However, as with any formula produced from experiments on humans, there are limitations; see the bibliography for comments on the formula's accuracy.

Loosely speaking, an individual's *maximum heart rate* (MHR) is the highest heart rate that can be sustained during prolonged exercise. Lucky for us, researchers have gotten many brave individuals to exercise at their MHRs in the name of science. These experiments have yielded equations that estimate MHR based on just age. You may already be familiar with the most popular formula: $MHR_{pop} = 220 - a$ (a linear function). But let's work out our quads instead—*quadratic polynomials*, that is—and use the formula with the smallest error [6]:

Estimating Your Maximum Heart Rate

$$MHR = 192 - 0.007a^2. \qquad (1.6)$$

Here MHR is your maximum heart rate (in bpm) and a is your age (in years).

This function is *quadratic* because the highest power of the independent variable (a in this case) is 2. Unlike linear functions, quadratic polynomials (and other nonlinear polynomials too) have "curvy" graphs.[11] Figure 1.2 shows a table and graph comparison of the two MHR formulas so you can see what I mean.

At Jason's age ($a = 23$) equation (1.6) says that his MHR is about 188 bpm, lower than the 192 bpm he would to achieve to burn off my snack. The linear MHR equation, however, disagrees—it says that Jason's *current* MHR is $220 - 23 = 197$. That's why the researchers responsible for (1.6) point out that "the traditional equation, $220 -$ age, *overestimates* [MHR] in young adults and *underestimates* it in older people" (emphasis original). That's exactly what Figure 1.2(b) shows, too (the linear graph is above the quadratic one until just before 40).[12] We conclude that Jason probably can't safely burn 400 calories in 20 minutes through aerobic exercise.

---

[11]If you're interested in learning more about quadratic (and higher-order) polynomials, see entry 6 in this chapter's appendix. A preview of what's there: linear functions are special cases of polynomials.

[12]Bonus: can you work out the exact age? See entry 7 in this chapter's appendix for the answer.

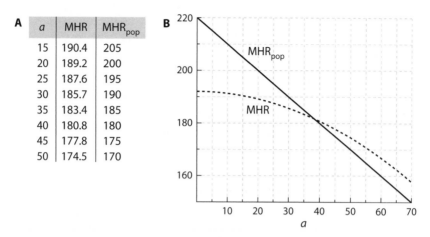

Figure 1.2. (a) A table of values of the quadratic function MHR (1.6) and the linear function $\text{MHR}_{\text{pop}} = 220 - a$. (b) The graphs of $\text{MHR} = 192 - 0.007a^2$ (dashed curve) and $\text{MHR}_{\text{pop}} = 220 - a$ (solid line).

Armed with (1.5) and (1.6), we can now answer other, more useful questions. Here's a classic one: if you exercise at $x\%$ of your MHR, how long will it take to burn c calories? This is such a common question, and I think it's neat we can now answer it using math.*8 Plus, you could follow the steps in the appendix and answer that question for yourself using your own versions of (1.5) and (1.6).

We now have a better understanding of precisely how RMR and aerobic exercise contribute to our *total daily energy expenditure* (TDEE). These are actually the two largest components, but there's one more contributor that's worth discussing: the "thermic effect" of food. In brief, this is the energy (calories) our bodies use up digesting food (this is additional energy not included in RMR). It turns out that certain foods require more energy to digest. I'll tell you which ones those are in the next section, and then we'll finally write down a formula for TDEE.

1.4 THE CALORIES REQUIRED TO DIGEST FOOD

The coffee shop I'm in probably gets its food shipments once a week. At those times, the employees expend energy moving the boxes out of

TABLE 1.2.
The thermic effect of each macronutrient and the net energy yield of ingesting 100 calories of each macronutrient [7, 8].

Macronutrient	Thermic Effect (% of Cal.)	Energy Yield (Cal.) of 100 Calories
Protein	20–35	65–80
Carbohydrate	5–15	85–95
Fat	3–15	85–97

the truck, unpacking them, and distributing the contents to the right places in the store. This is pretty much what our bodies do when we eat. That new shipment of calories requires unpacking (digestion) and distribution (absorption), and just like those coffee shop employees, our bodies expend energy (burn calories) doing all that. This additional energy needed to digest, absorb, and dispose of the food we eat is called *diet-induced thermogenesis* (DIT).

Each macronutrient has its own DIT effect. As Table 1.2 shows, proteins require the most energy to metabolize, followed by carbs and fats.

The thermic effect of food means, for example, that a 100-calorie snack *does not* provide your body with 100 calories—if the snack is pure protein, your body gets only 65 to 80 calories out of it, whereas if it's pure fat it gets 85 to 97 calories.

Since DIT is usually the smallest component of TDEE, and since we've now discussed the major ones, let's finally write down an approximate formula for TDEE:[13]

$$\text{TDEE} \approx \text{RMR} + \text{24-hour ACB} + \text{DIT}. \qquad (1.7)$$

(There are other factors that contribute to TDEE, but those three are the main ones.) Two quick comments: (1) the middle term here—24-hour ACB—accounts for all your day's aerobic exercise, and (2) since what you eat and how much you exercise varies day to day, *TDEE also varies day to day.*

Calculating TDEE each day can itself be a workout. That's why many fitness experts prefer to use a simpler formula.

---

[13] The $\approx$ symbol in the formula means "is approximately" (the Glossary of Mathematical Symbols in Appendix A explains this and other symbols used in the book).

> ### Estimating Your Total Daily Energy Expenditure
>
> $$\text{TDEE} \approx \text{RMR} \times \text{Activity Factor} + 0.1C. \qquad (1.8)$$
>
> Here TDEE is your total daily energy expenditure, RMR is the appropriate value from either (1.1) or (1.2), the Activity Factor is one of the values from Table 1.3, and C is your caloric intake for the day.

The "activity factor" in this formula describes how active your day was. Table 1.3 lists a few reference values.

After multiplying your RMR by the appropriate activity factor, (1.8) then adds 10% of your day's caloric intake to approximate the DIT term in (1.7).

Equations (1.7) and (1.8) were the end goals of this chapter. They give you a rough idea of your daily maintenance calories—the level needed to maintain your current weight. Health professionals tell us that eating fewer calories than our TDEE will create a *caloric deficit* and should lead to weight loss,[14] and that eating more calories than our TDEE will create a *caloric surplus* and should lead to weight gain. I've deliberately inserted the words "rough" and "should" because nutrition is not an exact science. From the Atwater general factor system's simplification of the energy yields of each macronutrient to

TABLE 1.3.
Activity factors; multiply your RMR by the appropriate factor to estimate your energy expenditure from RMR plus physical activity.

Level of Activity	Activity Factor
Little to no physical activity	1.2
Light-intensity exercise 1–3 days/week	1.4
Moderate-intensity exercise 3–5 days/week	1.5
Moderate- to vigorous-intensity exercise 6–7 days/week	1.7
Vigorous daily training	1.9

[14]But be careful not to spend too much time in a caloric deficit state—after a while your body adapts to the lower energy intake, a phenomenon known as *metabolic adaptation* [9]. This reduces (or may eliminate) your caloric deficit. Many of us know this outcome as the "plateau effect."

the approximations in (1.7)–(1.8), there's plenty of inaccuracy to go around. But there are ways to tame these errors. For example, after my Tuscan Fields shocker I capped my daily calories at my RMR and started eating more protein. The larger DIT effects of my protein-rich meals, along with the fact that I'm not a 24-hour couch potato, made it more likely that I was in a caloric deficit.

Our TDEE equations, and my response to the Tuscan Fields fiasco, illustrate perhaps the most important "equation" I can offer you:

$$\text{math} + \text{discipline-specific research-based knowledge} = \text{empowerment}. \tag{1.9}$$

Indeed, we've already built a solid foundation for healthy living by mathematizing the studies on nutrition and exercise—the "discipline-specific knowledge" part—presented in this chapter. In the next chapter I describe how the insights we've gained can help build a diet that improves cholesterol, lowers the risk of developing heart disease and diabetes, and may even add years to your life.

Chapter 1 Summary

MATHEMATICAL TAKEAWAYS

- Functions describe a special relationship between one or more variables, called the dependent variables, and one or more other variables called independent variables.
- We can visualize functions by graphing them. For a function with one dependent variable and one independent variable, each point on the graph is of the form (x, y), where x is the value of the independent variable and y the value of the dependent variable. The horizontal axis sets the scale for the x-values and the vertical axis sets the scale for the y-values.
- Linear functions are defined by two numbers: their slope and their y-intercept. The slope of a linear function can be interpreted as the change in the y-value resulting from a one-unit increase in the x-value. The y-intercept is the y-value of the point where the graph crosses the y-axis.

- Multilinear functions have multiple independent variables, each of which has an associated slope.
- The graphs of linear functions are lines. The graphs of nonlinear polynomial functions are curvy.

NONMATHEMATICAL TAKEAWAYS

- Nutrition labels state that carbs and proteins yield 4 calories per gram and fats 9 calories per gram, but this is a simplification (the Atwater general factor system).
- Linear functions—and their big brother, multilinear functions—show up frequently in nutrition science. The slopes of these functions help us better understand many fundamental concepts in nutrition (like the Atwater general factor system and RMR).
- You have a 24-hour calorie-burning machine: your body. Every day it burns your resting metabolic rate (RMR) in calories, which you can estimate using (1.1)–(1.2) and your weight, height, and age. A one-unit change in these numbers has a quantifiable effect on your RMR (Table 1.1). If you're male, the RMR equations predict that your body requires 166 more calories a day than a female of your same weight, height, and age.
- You can increase your daily energy expenditure through aerobic exercise, the most effective way to burn off body fat [4]. The additional calories you burn depends on how much heavy breathing you're doing: you burn about 5 calories for each liter of oxygen consumed. You can estimate your per-minute aerobic caloric burn (ACB) using (1.4) and your weight, age, and heart rate. Don't exert yourself too much, though—make sure to stay well below your maximum heart rate (1.6).
- Macronutrients require different amounts of energy (calories) to be metabolized, and this is measured by a macronutrient's diet-induced thermogenesis (DIT). Protein has the highest DIT; metabolizing it requires between 20% and 35% of the protein calories ingested. Table 1.2 summarizes the DIT of the other macronutrients.
- The sum of your RMR, your 24-hour ACB, and your meals' total DIT gives a rough estimate of your total daily energy expenditure (TDEE);

see (1.7). If your activity level and/or diet varies from day to day, your TDEE will also vary day to day.

- You can estimate your TDEE by using (1.8); see Table 1.3 for the appropriate activity factor to use.
- In theory, eating your TDEE in calories will help you maintain your current weight. Eating fewer calories than your TDEE will create a caloric deficit and may lead to weight loss, whereas eating more than your TDEE will create a caloric surplus and may lead to weight gain.
- Nutrition science is not exact; see the bibliography for brief comments on the errors and limitations of some of the formulas presented in this chapter.

BONUS: A FEW PRACTICAL TIPS

- *Get a pedometer.* A basic pedometer is cheap. It'll do a good job of counting the number of steps you take, and glancing at it often might motivate you to take more steps than normal (and hence burn more calories).
- *Track your calories with a smartphone app.* Apps make it easy to track your food intake and exercise. Some also allow you to set targets for your daily caloric intake, making it easier to stay close to your TDEE, and/or input your weight and/or body fat percentage.
- *Get a "smart" scale.* These days you can buy scales that measure your weight and body fat percentage; some even email you the results! One tip: weigh yourself at the same time and under the same conditions (e.g., before breakfast); this will cut out the weight fluctuations caused by eating and/or the time of day.

CHAPTER 2

■■■■■■■■■■■■■■■■

Live Longer (and Be Healthier) by Eating the Right Foods

AFTER I SWORE OFF TUSCAN FIELDS, I went looking for a new place to eat while I was on the go. I narrowed down my choices based on the nutritional information I found for each restaurant. Luckily, around that time I stumbled on a local place that served fresh food in customizable portion sizes and used high-quality ingredients (from a nutritional standpoint): Burrito Bob's. This Mexican-style restaurant sells burritos, quesadillas, and tacos. You pick what ingredients go in each (e.g., rice), which brings up a good question: what should you put in your burrito?

In this chapter we'll use this question to anchor our exploration of the health effects of individual macronutrients. Piggybacking on what we learned in the previous chapter, we'll use nutrition science to determine which burrito ingredients are healthy (and why). Afterword, I'll introduce you to a function that makes it easy to determine how healthy a particular food is. At the end of the chapter we'll talk about how the diet that emerges from these insights may even add years to your life (really). So let's pay Burrito Bob's a visit.

2.1 A GAME OF MACRONUTRIENT MUSICAL CHAIRS

Burrito Bob's simple menu is a bit deceptive—you actually have more choices than what's displayed. For example, you can skip the tortilla and opt instead for a "burrito bowl" (a burrito without the tortilla). That's exactly what I'm going to do. I've made my first choice already; let me explain why it's a healthy one.

2.1.1 Body Fat and Triglycerides: The Dynamic Duo

Of all the things you can put in a burrito at Burrito Bob's, the tortilla has the most calories (300 according to their nutrition pamphlet). My burrito bowl choice avoids those 300 calories right away. Plus, it avoids the end result of eating excess calories—increased body fat. (More precisely, excess calories are converted to *triglycerides*—a type of fat found in your blood—and triglycerides get stored in your fat cells [10].) But the devil's in the details. In this case that's the word *excess*. You see, what goes into body fat can also come out of body fat. In between meals, for example, your body draws on those triglycerides for energy. So body fat is not quite the static blob we envision it as; instead, it's more of a temporary store of energy.

So what's considered "excess"? Well, my old favorite—the 1,100-calorie fettuccine alfredo—clearly meets any definition of *excess*, yet the dynamic nature of body fat means that one meal's excess calories can be balanced out by another meal's low caloric content (e.g., soup for dinner). The solution is to define "excess calories" as *calories consumed above your TDEE* (i.e., the caloric surplus for the day).[1] Since excess calories increase body fat, we've now learned that *ending the day with a caloric surplus increases body fat*. So, a working knowledge of TDEE—and the multilinear functions behind it—helps with *weight* and *body fat* management.

The path to not increasing body fat is now clear: stay at or below your TDEE every day. But your body doesn't like it when you're below your TDEE. That forces it into a caloric deficit state, and since it has to make up the energy deficit, it burns some combination of two things for energy: your stored body fat (woo hoo!) and your muscle tissue (no!). Body fat just sloshes around as we move. But muscles are active tissues—they use energy as they contract and expand, and therefore they add to your overall caloric burn. That's why it's in your best interest to protect your muscle mass. If you don't and you're in a

---

[1] Studies suggest that 24-hour energy expenditure doesn't vary with meal frequency [12–14]. Participants in these studies ate between one and six times a day. The once-a-day-ers clearly ate an excess of calories for their one meal, yet their 24-hour energy expenditure was the same as the six-a-day-ers (whose meals had significantly fewer calories). This is why it makes more sense to define excess calories in terms of TDEE and not individual meals.

caloric deficit, the majority of your weight loss may end up coming from muscle mass. That would be terrible, since your body fat percentage—the ratio of your body fat weight to your total weight—would then *increase*; you'd have *lost* weight yet you'd look *fatter*!

Luckily there's one macronutrient that can protect you from this fate—our DIT all-star, protein. Let's talk about how much you need, and the unexpected benefits of increasing your protein intake.

2.1.2 Why Protein Packs a Punch

Let me discuss protein's magic by asking Andrea—the young woman taking my burrito bowl order—about Burrito Bob's chicken (she's wearing a T-shirt that says "Ask me about our chicken").

"Hi, I'd like a burrito bowl please. And, can you tell me about your chicken?"

"Sure," says Andrea. "One serving has 32 grams of protein, no carbs, and only 7 grams of fat; oh, and 180 calories."

"That sounds like a lot of protein."

"Yep, it's probably about half of your RDA."

"Say what?"

"Your Recommended Daily Allowance (RDA). For protein it's about 0.36 grams of protein per pound of body weight. I'd say you're 170 pounds, making your RDA for protein about 62 grams a day. That means one serving of our chicken is roughly half your protein RDA."

Wow, she knows her stuff! And how'd she guess my exact weight? "How did you know how much I weigh?"

"I'm also a personal trainer, so I can tell pretty easily how much someone weighs."

"Ah, makes sense. So, should I go for a double serving of the chicken to reach my RDA?"

"Probably not; you'll likely eat the remainder of your protein RDA throughout the rest of the day. If you're in a caloric deficit, though, you might want to double up the serving. In that state studies suggest you may need to consume up to 0.9 grams of protein per pound of body weight (per day) to prevent your body from cannibalizing your muscle for energy" [11].

I'm sure Andrea would have kept educating me on protein's benefits, but there were other people in line. What she said about protein is true, and the story gets even better—eating more protein also helps with weight loss! Here's what the authors of the DIT study I cited in Section 1.4 [7] have to say: "higher protein diets may facilitate weight loss when compared to a lower protein diet in the short term (within 6 months)." They cite the high DIT effect of protein as one explanation for this effect. For example, they calculate that burning a measly 40 calories more each day due to increased protein intake would lead to a weight loss of 4.2 pounds in a year. Plus, the authors report on another positive side effect of eating more protein: participants eating more protein reported "significantly increased subjective ratings of satiety" in 11 out of the 14 studies they reviewed. Translation: they felt fuller. That's quite a list of benefits for upping your protein intake!

But increasing protein intake without adding calories means *decreasing* either fat or carb intake. So, which should it be? I'll give you a clue: I just gave Eric—the guy next to Andrea—a "no thanks" when he asked me if I wanted rice in my burrito bowl. It turns out that a surprising number of healthy things happen when we cut back on our carb intake.

2.1.3 The Benefits of Reducing Carbohydrate Intake

Let me just hit you with it: "in controlled trials for weight loss, the LCD [low-carbohydrate diet, defined by the authors as between 50 and 150 grams of carbs per day] leads to weight loss and improvements in fasting triacylglycerols [triglycerides], HDL cholesterol, and the ratio of total to HDL cholesterol over a 6–12-month period." Amazing huh? That quote is from a recent review of nearly 100 studies on LCDs [16].[2] To really appreciate the significance of those findings, let's take a few minutes to discuss each one.

First up is the weight loss. Here's how the authors explain this side effect of LCDs: "the instruction to limit carbohydrate intake, without specific reference to calorie intake, leads to a spontaneous reduction

---

[2] Also, check out this nice summary of 23 randomized controlled trials studying the effects of LCDs [15] (the conclusions are similar).

in calorie intake." Makes sense; a common strategy for reducing carb intake is to eat more vegetables (which have fewer calories than the more popular carbs, like pasta).

What about the heart health claims? They sound too good to be true, right? Well, they aren't. A low-carb lifestyle can *dramatically* improve one's "lipid profile" (the blood test that checks cholesterol levels).

We've talked about triglycerides but not yet about cholesterol. You've probably heard the phrases "bad cholesterol" and "good cholesterol." The former refers to *low-density lipoprotein (LDL)*. Too much LDL floating around your blood could lead to plaque formation along your arteries and raise your risk for heart attack or stroke (hence the "bad" label). Luckily we have a built-in clean-up crew: *high-density lipoprotein (HDL)*. HDL particles take cholesterol in the blood stream back to the liver for disposal (that's why they get the "good" cholesterol label).[3] LDL and HDL are both lipoproteins, but as their names imply, they differ in their densities. "Densities *of what*?," you may wonder. Good question. To answer, that we need to go back to triglycerides.

You see, triglycerides can't be released directly into the blood stream. When your body needs to transport triglycerides through the blood (for example, in between meals when energy is being extracted from body fat) it packages them into lipoproteins. (This happens in the liver.) The lipoproteins produced are called *very-low-density lipoproteins (VLDLs)*. As VLDL molecules circulate in the blood stream, they lose triglycerides (your body draws on triglycerides for energy, remember?), decrease in size, and become more cholesterol-rich [18]. Eventually they morph into, you guessed it, LDL.

For decades scientists have known that LCDs lower triglyceride levels.[4] Now that you understand the connection between them and cholesterol, the heart-healthy properties of LCDs should make more

---

[3]Our understanding of cholesterol is continuously evolving. Total cholesterol was once considered the enemy; then we discovered LDL and HDL and they got labeled "bad" and "good," respectively. More recently we discovered that both LDL and HDL contain subparticles, some of which are more harmful than others. Thus, the bad/good distinction may be an oversimplification.

[4]In brief, *high* carbohydrate intake leads to high blood sugar, and one of the ways our bodies get rid of excess blood sugar is to store it as body fat (via the triglyceride to body fat pathway described in Section 2.1.1); see [19] for more details.

sense. What probably won't make sense is this: in many of the LCD studies, researchers replaced participants' carbs with *fat*. "What?!" I know. *Adding* fat to our diet to *improve* our health—and heart health in particular—is so counterintuitive that it merits discussion, so that's our next stop.

2.1.4 Dietary Fat: Friend or Foe?

Let's go back to the review study [16] that cited those heart-healthy effects of low-carb diets. Table 3 of that publication details the macronutrient breakdown of the low-carb diets included in the review. The kicker? *No diet had less than about 40% fat intake.* Eat 40% of your calories as fats and improve your heart health? That sounds crazy!

Before you call 'em crazy, let's turn to the research. A recent *meta-analysis*—a study of studies—of 60 controlled trials [20] studied the effect on cholesterol of swapping carbs for fats in your diet. The authors found that swapping carbs for *unsaturated* fats *improved* cholesterol numbers, while swapping carbs for *saturated* fats generally *worsened* cholesterol numbers. They also came up with an equation that predicts the total-to-HDL cholesterol ratio (hereafter THR) after the swap [20, Figure 2]:[5]

Estimating Your THR after Swapping Carbs for Fats

$$THR = 0.003s - 0.026m - 0.032p + b. \qquad (2.1)$$

Here THR is the total-to-HDL cholesterol ratio. The other variables' meanings (and the error ranges of the coefficients) are discussed below.

---

[5]The authors cite studies showing that THR is a better indicator of heart disease risk than total cholesterol or individual cholesterol (e.g., LDL) levels. Lower THR implies a lower risk for heart disease. Also, see the bibliography for a discussion of the study's limitations.

TABLE 2.1.
The predicted effect on the total–to–HDL cholesterol ratio (THR) of swapping 1% of carbohydrate calories for fat calories. See the paragraphs below equation (2.1) for a discussion of the error ranges for each coefficient.

If you swap 1% of carb calories for...	Your total-to-HDL ratio goes...
Saturated fat calories	Up by 0.003
Monounsaturated fat calories	Down by 0.026
Polyunsaturated fat calories	Down by 0.032

Here's what each term means.

- **The variables s, m, and p.** The variable s represents the percentage of carbohydrate calories swapped for *saturated* fat calories. (For example, $s = 5$ indicates you're replacing 5% of your carb calories with saturated fat calories while keeping the rest of your diet the same.) The m and p variables have similar interpretations: m is the percentage of carb calories swapped for *monounsaturated* fat calories and p the percentage of carb calories swapped for *polyunsaturated* fat calories.
- **The meaning of b.** The number b is your current THR ratio (i.e., before swapping any carb calories for fat calories); the authors describe b as the "intrinsic total–to–HDL concentration."

Now (2.1) is a multilinear function, just like the RMR equations of the last chapter. So, we can use the same approach to interpret the coefficients of the variables in (2.1). Table 2.1 summarizes those interpretations.

Notice that swapping 1% of your carb calories for *saturated* fat calories is predicted to *increase* THR by 0.003. (That's bad, since higher THR values correlate with an increased risk for heart disease.) But there's a catch. You see, each coefficient in (2.1) has an error range (see Table 1 in [20]). The ranges for the m and p coefficients only include negative numbers, but the range for the s coefficient includes positive and negative numbers. That means the researchers' analysis doesn't reliably predict the effect of swapping 1% of your carb calories for saturated fat calories (i.e., *increase* or *decrease* THR).[6] Swap the carbs

---

[6]There is growing evidence to support the claim that saturated fats *don't* cause heart disease; see [21] for a list of recent studies examining this issue, or the (easier to read) June 23, 2014, issue of *Time* magazine, whose cover read "Eat Butter. Scientists labeled fat the enemy. Why they were wrong."

for *un*saturated fats, however, and even after accounting for the error ranges, (2.1) still predicts a *decrease* in THR (and therefore in your risk for developing heart disease). See, multilinear functions are also good for your heart!

Another takeaway from this analysis is that some fats are *good* for your heart.[7] That's why I just asked Eric, the Burrito Bob's guy finishing up my order, to add a serving of avocado to my burrito bowl. One cup of avocado has 15 grams of monounsaturated fat and only 3 grams of saturated fat. And because I skipped the rice earlier, adding the avocado is like substituting carbs for (mostly) monounsaturated fat, which lowers my THR via the -0.026 coefficient in (2.1).

It's tempting to take these findings to the extreme and replace *all* carb intake with unsaturated fats. But just as there are good fats, there are also "good carbs." I'll tell you about the star of that show in the next section.

2.1.5 Fiber: The Unsung Hero

Dietary fiber is broken down into two forms on the back of nutrition labels—insoluble and soluble. Fiber is typically described as "indigestible," and that's largely true (more on that in a minute). "What's the point of eating something you can't digest?," you might wonder. Well, it turns out that fiber has amazing health benefits. Insoluble fiber, for example, passes straight through your digestive system, adding bulk, and therefore helping make you more regular. But soluble fiber is the real rock star.

Soluble fiber dissolves in water ("soluble") to form a gel, and as this gel travels through your small intestine, it scoops up LDL cholesterol [23]. When your dinner comes out the other end, that gel—and the cholesterol bound to it—is removed. Presto! Lower circulating cholesterol. This partly explains why people who eat lots of fiber have lower risks for all kinds of diseases, including heart disease and diabetes [22]. If you eat black beans, this is good news—one cup contains about

---

[7] In Table 2 of [20] the authors examine individual saturated fats and find that lauric acid, found in high concentrations in coconut oil, lowers THR. It seems, therefore, that even some types of *saturated* fats are good for you.

4 grams of soluble fiber, virtually no fat, and 15 grams of protein [24], and eating just *half a cup* a day for 3 weeks is enough to lower LDL cholesterol by 20 mg/dl [25].

Another hidden benefit of swapping some of your carbs for fiber is the accompanying calorie reduction. "Net carb" proponents claim that because fiber is indigestible it yields 0 calories per gram; therefore, the argument goes, one should subtract the grams of fiber from the total carbohydrate number (giving the "net carb" number) and recalculate the food's calorie content. But fiber can be fermented by bacteria in the colon. In some people this releases energy the body can use [26]. This person-to-person variability has made it difficult for scientists to assign an Atwater factor to fiber, but a generally accepted value is 2 calories per gram [17]. So, feel free to subtract two times the fiber content from the total calorie count of a fiber-rich food.[8]

Let me mention one last benefit of eating more fiber. Say you were trying to limit your carbs to 150 grams per day. As we just discussed, you could further decrease the accessible carb calories—and hence create a larger caloric deficit—by eating a significant portion of your 150-gram allotment as fiber. But viewed another way, you could eat much more food if the majority of those calories came from fiber-rich foods. That's why I referred to fiber as the unsung hero—replacing your carbs with more fiber-full ones is a heart-healthy choice, helps you become more regular, and either helps stealthily decrease your caloric intake or helps you keep caloric intake constant but eat a larger quantity of food. That's also why I asked Eric to dump a healthy portion of black beans into my burrito bowl to finish up the order.

This section's given you a lot to digest. But don't worry, I'll summarize it all in the chapter summary. For now, let me reiterate two foundations for a healthy life math has led us to: being mindful of your TDEE and understanding the health effects of each macronutrient.

---

[8]For example, 1 cup of black beans contains 227 calories and 15 grams of fiber. The fiber-adjusted calorie count is 197 (227 minus the 2(15) = 30 from the fiber adjustment). Read the nutrition labels carefully though—the food's calorie count might already have been adjusted for the fiber content (FDA regulations give food manufacturers the option to subtract [or not] the fiber calories from the total calories).

TABLE 2.2.
Comparison of the energy density of the ingredients in my burrito bowl (nutritional information from nutritiondata.com).

Food	Calories per Serving	Weight per Serving (grams)	Energy Density (calories/gram)
Chicken	180	113	$\frac{180}{113} \approx 1.6$
Black beans	120	113	$\frac{120}{113} \approx 1.1$
Pico de gallo Salsa	20	99	$\frac{20}{99} \approx 0.2$
Cheese	100	28	$\frac{100}{28} \approx 3.6$
Guacamole	230	113	$\frac{230}{113} \approx 2$

The holy grail remains, though: is there a simple way to choose foods that help you stay under your TDEE and give you the health benefits we've discussed? Yes. And it'll only take *one* function, as we'll see in the next section.

2.2 HOW TO EAT MORE AND BE HEALTHIER: ENERGY DENSITY

Your TDEE is your daily food budget. Just like with any budget, we'd like to get the most bang for our buck. That's where **energy density** comes in.

The concept is simple: divide the calories in a food item by the weight of the item. Let me illustrate the concept by calculating the energy densities of the ingredients in my burrito bowl (see Table 2.2).

The first thing to mention is that our "calories per gram" calculation may not be accurate for foods with fiber. If the fiber calories have already been subtracted from the total calories, we're okay. Otherwise, subtracting those fiber calories will decrease a food's energy density. Take the black beans, for example. The 120 calories in those beans includes the calories that come from the 12 grams of fiber the serving contains. Removing 24 calories—recall the 2 cal/g energy yield assigned to fiber—lowers the energy density to about 0.85 calories per gram. Similarly, the avocado in the guacamole also contains fiber, so the guacamole's actual energy density is lower than the 2 cal/g calculated in the table.

Okay on to the fun part—let's look for patterns. Notice that almost everything has an energy density of at most 2 calories per gram. The one thing that doesn't is the saturated-fat-heavy cheese. So, we've found a possible "threshold" (2 cal/g) that divides the healthy ingredients in my burrito bowl from the unhealthy one. But how general is this 2 cal/g threshold? If we were to calculate the energy densities of other common foods and assemble them into a table, would 2 cal/g still roughly divide the healthy foods from the unhealthy ones? Yep, just about, as the British Nutrition Foundation's *Feed Yourself Fuller* chart illustrates (Figure 2.1).

The first column of the chart is energy density paradise—fruits and vegetables give us great bang for our calorie buck. Some of my Burrito Bob's favorites (beans and chicken) appear in the second column, along with other staples like yogurt and eggs. The third column is the gray area. From "jam" and "meat pizza" onward, we know we're in trouble, but before that, I'd consider a lean steak and grilled salmon to be healthy options (eaten in moderation), but wouldn't touch the chocolate mousse. The last column has the worst bang for calorie buck foods—cooking oils, processed carbs (like crackers), and high-fat additives like butter. So it seems that 2 cal/g does do a good job of separating the good from the bad!

"Okay, but where's that holy grail function you promised?" Glad you asked. Let's get there by thinking about things differently. Let's say you have 100 calories to "spend." What foods will fill you up the most?

To illustrate this new approach, let's look at Figure 2.1 and compare the banana to the croissant. First, note that the croissant is almost four times as energy dense as the banana ($3.7/0.95 \approx 3.9$). Translation: for the same amount of calories we can eat four times as many bananas as croissants. For example, for the same 100 calories we can eat about 105 grams of bananas but only about 27 grams of a croissant.[*1] That means the bananas will fill you up much faster than will the croissants

Figure 2.1. Excerpt from the British Nutrition Foundation's *Feed Yourself Fuller* chart, available as a free download [27]. © British Nutrition Foundation, 2010.

(they'll literally make your stomach feel four times as heavy as croissants will). Plus, there's also the visual effect: for the same amount of calories, I'd rather eat one medium-sized banana—which would take me quite a few bites—than one mini-croissant (you know, the ones about the size of a computer mouse), which I could eat in two bites.

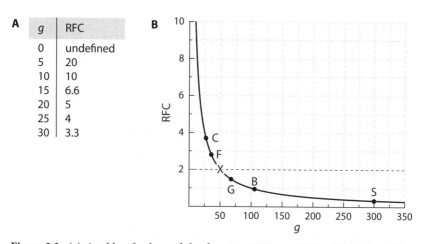

Figure 2.2. (a) A table of values of the function RFC = 100/g. (b) A plot of the function. The points correspond to the energy densities of bananas (B), croissants (C), grilled chicken (G), French fries (F), and strawberries (S). The X marks the intersection of the curve with the line $y = 2$.

This new approach suggests that energy density becomes a more useful concept if we imagine eating different foods that all contain the same amount of calories. So, without further ado, let me introduce you to what I'll call the *rational food choice* function:

$$\text{RFC} = \frac{100}{g}. \tag{2.2}$$

Here, g is how much a food weighs, measured in grams, and the numerator (100) has units of calories. Thus, (2.2) has units of calories per gram (energy density).

Mathematically, the RFC differs from the other functions we've encountered; it's a simple example of a *rational function*—a function of the form "polynomial divided by polynomial." The presence of division means that a rational function may be undefined for some values of the independent variable (we can't divide by zero; here's why[*2]). (For example, the RFC function is not defined at $g = 0$.) The graphs of rational functions often shoot off to infinity as we approach the values where the function isn't defined. This happens to the RFC function as g gets closer to zero (see Figure 2.2(b)).

Figure 2.2(b) contains important nutritional insight. For starters, note that different foods will appear along different places on the RFC graph according to their energy densities. Each food's y-value is its energy density, and its x-value tells us how many grams of it we have to eat to reach 100 calories. For instance, point C (for croissant) has a y-value of 3.7 and an x-value of about 27 (which we calculated earlier), meaning that a croissant has an energy density of 3.7 cal/g and you need to eat about 27 grams to get 100 calories' worth. By contrast, check out point S. Don't think swapping one mini-croissant for one banana will fill you up? How about swapping that croissant for *92 strawberries* (about 300 grams by weight)? The takeaway: *the farther right on the RFC curve the food appears the more of it you can eat (weight-wise) for the same 100 calories.* Those foods will make you feel fuller than foods on the left side of the curve. And since you'll be too full to keep eating, that'll help keep your caloric intake low.

But the RFC function contains yet more insight. Were we to plot more points, its graph would eventually contain all the foods in Figure 2.1. That means it would also contain the magic 2 cal/g threshold we settled on. Even with just the few foods I plotted from Figure 2.1, notice how the thickest dotted line at the y-value of 2—corresponding to our magic 2 cal/g energy density—visually divides the graph into a region of rapidly rising energy density (to the left of the X in the graph) and a region of steadily changing energy density (to the right of the X). The implication is that you shouldn't worry about swapping foods within the right region—since one food's energy density may only be slightly higher than the other—but you should be careful swapping foods in the left region since you might inadvertently add more calories for the same weight of food ingested. Finally, notice that the right region is the energy density paradise of Figure 2.1. That's where you'll find the fruits, vegetables, lean meats, unsaturated fats, and fiber. These are the foods that make us feel full without having to eat too much, help us maintain (or lose) our weight, and offer us protection against the major diseases.

Rational functions aren't the usual heroes of stories. But the RFC function is the hero of this one; it's your one-stop shop for healthy living. In the next section I'll continue our tour of functions and bring back some familiar friends (polynomials) to help add years to your life.

2.3 LIVE LONG(ER) AND PROSPER WITH THE WAIST-HEIGHT RATIO

In 2005 two researchers published a paper titled "Six reasons why the waist-to-height ratio is a rapid and effective global indicator for health risks of obesity and how its use could simplify the international public health message on obesity" [28]. It's a long title, but it says it all. The researchers wrote about mounting evidence supporting the usage of the *waist-height ratio* (shortened to WHtR) as a predictor of the risk for developing cardiovascular disease and diabetes.[9] A subsequent systematic review of the research literature found that a WHtR larger than 0.5 *increases* your risk for developing those diseases [29]. It isn't surprising that people with those diseases live shorter lives. What is surprising is that researchers have been able to quantify the loss in life span based on the WHtR value.

One study that did this was published in 2014. Researchers used 20 years' worth of health data on the British population (England, Wales, and Scotland) to calculate the "years of life lost" (YLL) based on individuals' WHtR [30]. The paper includes several tables and figures that showed how many years of life female and male nonsmokers age 30, 50, or 70 lost when their WHtR increased above 0.5. (Subsequent studies have found similar results for non-British populations.)

But what if you're not 30, 50, or 70? Wouldn't it be nice to have an equation and not have to squint to estimate the YLL from their tables? That's where math can help.

First, let's denote the WHtR by r, the male nonsmokers' YLL by y_m, and the female nonsmokers' YLL by y_f. Next we add subscripts to the y's that keep track of the three age groups included in the study (for example, the YLL by 30-year-old male nonsmokers will be denoted $y_{m,30}$). With all that said, here are the equations for 30-year-old males and females (I've put the rest of the equations in the appendix):*[3]

$$y_{m,30} = 616.67r^3 - 920r^2 + 467.83r - 81,$$

$$y_{f,30} = 150r^3 - 175r^2 + 69r - 9.4.$$

---

[9]The WHtR is calculated by taking the ratio of your waist circumference—the distance around your waist—to your height (where both are measured in the same units, e.g., centimeters).

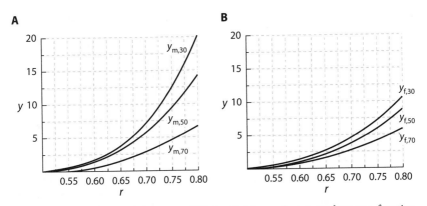

Figure 2.3. (a) Graphs of the years of life lost (y) for male nonsmokers as a function of their waist-to-height ratio (r) for three age groups: 30, 50, and 70. (b) Similar graphs for female nonsmokers. See the appendix for the equations of these curves.[*3]

These equations—and the rest in the appendix—are the ugliest we've seen yet. But at least their graphs are pretty. Figure 2.3 shows the graphs of all six YLL equations. Let me point out a few key observations.

- **Men lose more years of life than women for the same WHtR value.** We see this by directly comparing curves for the same age group, for example, $y_{m,30}$ and $y_{f,30}$. (I've kept the x- and y-axes' scaling the same in both graphs, so it's okay to directly compare the curves.) For example, the life span of a 30-year-old man with a WHtR of 0.8 is 20 years shorter than his life span with a WHtR value of 0.5, but for a 30-year-old woman with the same WHtR of 0.8, only about 11 years of life are lost relative to her 0.5 WHtR self.

- **More years of life are lost when WHtR increases from an already high number vs. a lower number.** For example, have a look at $y_{m,30}$ in graph (a). From $r = 0.7$ to $r = 0.75$ the YLL increases by 5. However, from $r = 0.75$ to $r = 0.8$ the YLL increases by 7.5. The takeaway: even a small increase from a high WHtR can lead to a significant increase in YLL. (These differential effects are happening because the equations are nonlinear.) I'm an optimist, so here's that statement in "glass half full" form: *even a small decrease from a high WHtR can lead to a significant decrease in YLL.* Case in point: a 30-year-old male nonsmoker with a WHtR of 0.8 who reduces that number to 0.75—only about a 6% reduction—gains back 7.5 years of life!

- **The effect of WHtR on YLL diminishes with age**. For example, if you're a 70-year-old woman with a WHtR of 0.8, your YLL is roughly half that of a 30-year-old woman with the same WHtR. This may be partially due to *survivorship bias* (i.e., perhaps some of the 30-year-old women with very high WHtR values didn't make it to age 70).
- **A WHtR of 0.5 minimizes YLL**. We already mentioned this. But it's nice to see all graphs in Figure 2.3 reflect this (all curves' *y*-values are nearly zero when $r = 0.5$).

"Very informative," I hope you're thinking, "but what if I'm not 30, 50, or 70?" Well we're in mathland now. And since we now have YLL *equations*, we're no longer bound by the data on the three age groups reported in [30]—we can use our equations and math to estimate YLL for any age (see the appendix).*4

So, how do you gain back the years of life you're projected to lose given your current WHtR? Well, if you're done growing, then your height won't change, so you've got to decrease your waist circumference. I know what some of you are thinking: liposuction! But I doubt it's that easy. A more beneficial approach is to lose that belly fat through better nutrition and exercise. That brings us full circle to the first page of the book—the search for a healthy diet. The one we built from the ground up by applying math to the research on nutrition—the low-carb, higher protein, high fiber, good dietary fat, low energy density, and TDEE-mindful lifestyle—now gets another claim to fame: it may help you live longer.[10] Score another (huge) one for math!

I hope I'm convincing you of the powerful and everyday applications of mathematics. And we're just getting started—the insights we've gained in these past two chapters are the proverbial tip of the iceberg. Read on and you'll learn how math can help you retire early and even find love. These seem like grandiose claims, but here's what

---

[10]Recently, the Dietary Guidelines Advisory Committee—a panel of nutrition experts that informs the government's Dietary Guidelines for Americans—reviewed the scientific evidence on health and nutrition and concluded (drum roll, please) ... that we should limit saturated fat intake, eat more polyunsaturated fat (both consistent with (2.1)), not worry about limiting the amount of total fat in the diet (!), limit sugar intake, eat more whole grains, eat more fruits and vegetables, and exercise more [31, 32].

one scientist had to say about how surprisingly applicable math is [33]: "How can it be that mathematics, being after all a product of human thought which is independent of experience, is so admirably appropriate to the objects of reality?" (Albert Einstein).

Chapter 2 Summary

MATHEMATICAL TAKEAWAYS

- Rational functions are functions of the form "polynomial divided by polynomial." Their graphs may sometimes shoot off to infinity, which partly results from trying to divide by zero.
- In nonlinear polynomials (like the cubic polynomials in Figure 2.3) a one-unit increase in the x-value has different effects on the y-value depending on the starting x-value (see the second observation regarding Figure 2.3). This stands in contrast to what happens with linear functions (in which a one-unit increase in x *always* adds or subtracts the same number from y—the slope).

NONMATHEMATICAL TAKEAWAYS

- The excess calories you eat are packaged into triglycerides and stored in your fat cells; thus, excess calories increase body fat. Between meals triglycerides are pulled from your fat stores and later burned for energy. These facts imply that body fat is less of a static entity and more of a temporary store of energy.
- Eating at most your TDEE in calories should prevent you from gaining body fat.
- Upping your protein intake will increase your daily energy expenditure via its high DIT effect. It'll also make you feel fuller and help you preserve muscle mass if you're in a caloric deficit. (Even better, strength train a few times a week to avoid losing muscle mass.)
- Research shows that low-carbohydrate diets improve various markers of disease over the short term, including high cholesterol and triglyceride levels. In addition, low-carb diets may also lead to weight (and body fat) loss in the short term.

- Swapping carbs for an equal amount (calorie-wise) of unsaturated fats has been shown to reduce the total-to-HDL ratio and therefore lower your risk for developing heart disease. (See (2.1) and Table 2.1.)
- Don't cut out the carbs entirely—replacing them with fiber-rich carbs (like black beans) has been shown to lower LDL cholesterol, which reduces your risk for developing heart disease. Eating more fiber will also make you more regular and help cut your caloric intake.
- Eating low energy density foods will help you feel fuller on fewer calories, thereby helping your caloric intake stay around your TDEE. As you calculate the energy densities of the foods you eat, plot the corresponding points on the RFC function (2.2); this will give you a visual way to improve your diet (eat more of the foods appearing to the right of the "X" and less of those appearing to the left of it).
- An energy density of roughly 2 calories per gram separates healthy foods from unhealthy ones. To apply this rule of thumb on the go, simply look at a food's nutrition label and divide the calories in one serving by the grams in the serving.
- Keeping your waist-to-height ratio (your waist circumference divided by your height) as close as possible to 0.5 can help you live longer. You can estimate the years of life lost based on your current waist-to-height ratio using the equations in the appendix.[*3]

BONUS: A FEW PRACTICAL TIPS

- *Dedicate each week to replacing a different food item in your kitchen with a healthier alternative.* I used to love white bread. Now I've gotten used to eating fiber-rich whole grain bread. It took several trips to several markets to find a good bread, but it paid off. Ditto for chocolate (very dark chocolate has a lot of fiber and other health benefits).
- *Change your diet slowly and use intermediaries.* I used to love soda (pure sugar) and drank little to no water. These days I love water and drink little to no soda. But that didn't happen overnight. I first swapped the soda for diet soda; this cut out the calories yet kept much of the soda taste. Then, once I got used to that I swapped the diet soda for sparkling water; this cut out much of the artificial ingredients but

kept the fizz. The move to plain water was then a lot easier than had I gone from soda to water directly.

- *Get a food scale.* Spend a few days weighing the food you consume to better understand your current portion sizes. This will also help you more accurately estimate the calories you're eating.
- *Find an app that aggregates fast food nutrition information.* There are many free apps that contain the nutritional information for the most popular fast food places. Some of these apps even have a search feature; you can use it to search a restaurant's menu, for example, for options with fewer than 100 calories.
- *Use the nutritiondata.com website.* I haven't found a better free source of nutrition information that's as detailed and user-friendly as this website. You'll find the basics—nutrition information for tons of foods—and a highly customizable search interface (for example, you can search for foods highest in fiber and lowest in calories), and very detailed information on a food's nutrients.

PART II

A Mathematician's Guide to Managing Your Money

CHAPTER 3

■ ■ ■ ■ ■ ■ ■ ■ ■ ■ ■ ■ ■ ■ ■ ■

Dissecting Your Monthly Budget

IN THE LAST CHAPTER I encouraged you to think of your TDEE as your daily food budget and view your food choices as opportunities to spend those calories. I introduced you to the concept of energy density and we then searched for foods that give us the most bang for our calorie buck. "Budget," "spend," "bang for buck"—these are phrases we usually associate with money. Could there be a connection between nutrition and money? *Yes.* I deliberately planted those words in the chapters about nutrition to foreshadow our discussion of that connection. Think about it—TDEE is like your monthly budget, and calories (the currency of TDEE) are like money (the currency of the monthly budget). Further, spending more than your monthly budget (a "budget deficit" state) can have negative consequences, just like eating more than your TDEE (a caloric surplus state).

In this part of the book we exploit this nutrition-money connection to discuss personal finance using the same approach I took to explore personal health in the last two chapters. I'll dissect our finances and use math to help uncover useful insights into each component we run into. We'll focus on your monthly budget in this chapter and on the broader economy and investing in the next chapter. We will start off by breaking down the monthly budget into income, taxes, and expenses, and will meet some new functions—exponentials and logarithms—that will help us better understand these components. We'll discover that they can help us increase our income, cut our taxes and expenses, and help us make better financial decisions. Math will even give us what *Lord of The Rings* characters might call "the one equation to rule them all": an equation that determines when we'll reach financial independence, that liberating date when you can quit your job and live off your savings alone. Pretty powerful stuff. Let's get started.

3.1 THE RETURN OF THE KING (THE LINEAR FUNCTION)

Let's kick off this new adventure at a familiar location—your neighborhood. Somewhere inside your home, perhaps on your kitchen table, are this month's bills. In Chapter 2 we learned that different sources of calories affect our bodies in different ways (recall the different DIT effects and short-term health implications of each macronutrient), and as you sift through your bills you'll notice that money works the same way. Some types of money *subtract* from your monthly budget (for example, debts), and other kinds *add* (for example, interest on savings). To sort out the good money from the bad money, let's start at the heart of the monthly budget: income.

To keep things concrete, let's pretend that your only source of income is a job that pays $10 an hour.[1] If you work x hours in one year, your earnings E is a linear function of x:

$$E = 10x. \tag{3.1}$$

The slope of 10 tells us that for every additional hour you work, you'll earn $10. But keep in mind that (3.1) is your *gross income*—income before taxes are taken out.

Taxes? Taxes! I know, but don't panic. Taxes boil down to math, and that means we can use what you've already learned in this book to understand (and lower!) your tax bill. Here's how.

Let's focus on your federal taxes—taxes paid to the federal government—for the time being. (We'll talk briefly about other taxes

---

[1]You can customize everything we're about to do by replacing $10 with your actual hourly earnings.

TABLE 3.1.
The 2015 IRS tax brackets and tax due by taxable income range (for "single" status filers). Source: irs.gov.

Yearly Taxable Income	Tax Due	Tax Bracket
$0–$9,225	10% of taxable income	10%
$9,225–$37,450	$922.50 + 15% of the amount over $9,225	15%
$37,450–$90,750	$5,156.25 + 25% of the amount over $37,450	25%
$90,750–$189,300	$18,481.25 + 28% of the amount over $90,750	28%

later.) Those are calculated according to the following simple three-step process:

1. Calculate your *taxable income*:

    Taxable income = Gross income − Allowed deductions.

2. Determine your *tax bracket* based on your taxable income.
3. Use that bracket's formula to calculate your taxes due.

That's it! Let's talk about each piece and what it means for your finances.

The first thing to know is that the United States, like many other countries, has a *progressive income tax*—high-income earners pay a higher percentage of their earnings in taxes than lower-income earners do. I'll keep things simple and assume that you file your federal return as a "single" taxpayer and that you can't be claimed as someone else's dependent. Table 3.1 shows the first few tax brackets for 2015 set by the Internal Revenue Service (the tax collector) in that case. (There are more tax brackets than those shown in Table 3.1; the next one starts at $189,300 and goes up to $411,500.) Notice that the tax bracket is identified by the percentage appearing in the "Tax Due" column. These percentages also determine how much of each dollar you earn you get to keep. To see how this goes, we need to mathematize Table 3.1.

Let's start with the first tax bracket. Let's denote your taxable income by z and your tax due by T. The first line of Table 3.1 then says that if your taxable income is between $0 and $9,225 (i.e. $0 \leq z \leq 9{,}225$) then the tax due is 10% of z:

$$T = 0.10z, \quad 0 \leq z \leq 9{,}225. \quad (3.2)$$

Hey, it's another linear function! This time the slope is 0.10, which tells us that in the 10% tax bracket, each additional dollar you earn increases the tax due by 10¢.[2] Thus, you keep only 90¢ of each dollar you earn in the 10% tax bracket.

Let's see what happens if your taxable income is between \$9,225 and \$37,450 (i.e., $9,225 \le z \le 37,450$). The table says you're now in the 15% tax bracket, and the tax due is[*1]

$$T = 922.50 + 0.15(z - 9,225), \quad 9,225 < z \le 37,450. \quad (3.3)$$

This linear function has slope 0.15, so each additional dollar earned gets 15¢ in taxes taken away (5¢ more than in the 10% bracket). Conclusion: in the 15% tax bracket you keep only 85¢ of every dollar earned.

Were we to continue mathematizing each line of Table 3.1, we would end up with four linear functions, each corresponding to a different taxable income range. These four functions can be combined into one function—a *piecewise linear* function (a function made up of pieces of linear functions):

Calculating Your Federal Tax Due

$$T = \begin{cases} 0.10z, & 0 \le z \le 9,225, \\ 922.50 + 0.15(z - 9,225), & 9,225 \le z \le 37,450, \\ 5,156.25 + 0.25(z - 37,450), & 37,450 \le z \le 90,750, \\ 18,481.25 + 0.28(z - 90,750), & 90,750 \le z \le 189,300. \end{cases} \quad (3.4)$$

Here T is your 2015 federal tax due and z is your taxable income— gross income minus tax deductions. This assumes "single" filing status.

---

[2]Technically (3.2) says that for each additional dollar of *taxable income* earned the tax due increases by 10¢. I assume hereafter that your deductions (which we discuss shortly) stay the same, so that increases in taxable income are the same as increases in gross income.

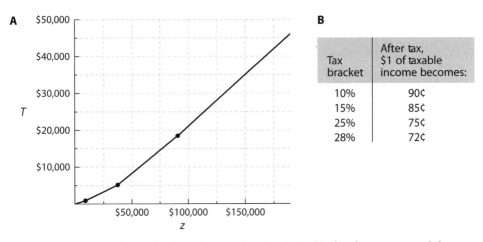

Figure 3.1. (a) The graph of the piecewise function (3.4). (b) The after-tax amount left from earning $1, by tax bracket.

To actually use (3.4) you first determine which of the linear functions on the right-hand side to use. You figure that by knowing the value of the independent variable (z in this case). For example, if your taxable income were $10,000 we'd use the second linear function in (3.4), since $z = \$10,000$ lies in the second range of numbers.

Now that we've mathematized the IRS tax table, let's see what the math can teach us. First, let's get a bird's-eye view of the action. Figure 3.1(a) shows the graph of the piecewise function (3.4) (the dots separate the different linear function pieces). The accompanying table confirms what the math has already hinted at: your tax bracket determines how much of each dollar of taxable income earned you get to keep.

We can use this insight to help increase our monthly *after-tax income*, sometimes also called *take-home pay*, or to an economist, *disposable income*:

$$\text{Disposable income} = \text{Gross income} - \text{Income taxes.} \qquad (3.5)$$

According to this formula, increases in disposable income come from either increasing gross income or decreasing income taxes (or both). I'll share some simple ways to generate additional income in the chapter summary. For now let's go back to (3.4) with a critical eye to see if it can tell us how to lower our taxes (and therefore increase our take-home pay).

Look closely at the formulas on the right-hand side of (3.4). They all depend on the same variable: z—your taxable income. So, the most straightforward way to lower your federal tax bill is to decrease your taxable income. But remember, taxable income is gross income *minus* allowed deductions. We don't want to lower our gross income, so that leaves one clear target: *deductions*.

Deductions are amounts the government doesn't tax you on. There are *many* allowed deductions, but to keep things concrete let's assume you only deduct the "standard deduction" ($6,300 in 2015 for a single filer) and one "personal exemption" ($4,000 in 2015 for a single filer). Your taxable income equation is then

$$z = 10x - 10,300. \tag{3.6}$$

Notice that since the first $10,300 (the sum of your two deductions) of your earnings don't get taxed, this means you could work 1,030 hours and *not pay federal tax*.[3] That works out to about 20 hours per week—a part-time job.

What if you want to work *full*-time *and* lower your tax bill? Well if you increase your deductions we subtract more than $10,300 in (3.6) and you can work more hours before you have to start paying federal tax. This is why the most important question come tax time is always: have you deducted everything you possibly can?[4] There are even some deductions that the government *encourages* you to take. For example, contributions to certain retirement accounts and interest paid on student loans and mortgages are all deductible.

Even if you claim every deduction and credit you qualify for, you still have other taxes to pay; for example, in the United States that includes payroll taxes (which include Social Security and Medicare taxes) and

---

[3]This number comes from solving $10x - 10,300 = 0$, which gives $x = 1,030$.

[4]A related important question is: have you taken all the tax *credits* you qualify for? Deductions make z (taxable income) smaller, but tax credits make T (tax due) smaller. That's why tax experts say that tax credits reduce tax due "dollar for dollar."

state taxes. These serve important purposes—like funding schools—but it can get difficult to keep track of them all.[5]

Okay, let's pause for a moment and pat ourselves on the back. We've scored a victory here. We took a complicated task—making sense of your federal taxes—and made it simpler by using math. The equations that arose helped us understand how much of each dollar we earn we get to keep. Plus, by thinking carefully about the taxable income equation, we came up with strategies to *reduce* our tax burden. I hope this reminds you of (1.9)—the "math $\rightarrow$ empowerment" equation; though we've only scratched the surface, it should be clear that math can even conquer taxes!

After you flex your new, bigger, tax-crushing muscles, you'll be left thinking about your disposable income again. It would be nice if you could "dispose" of this money as you wish. But certain things—like bills—need to be paid first. Here again math has much to say. The insights, however, depend on the expense. So let's talk expenses.

3.2 TO EXPENSES, AND BEYOND!

The term *bills* is too broad a term for our purposes. Go back to your own stack of bills and you'll see that the majority fall into one of two categories: *necessary expenses* (e.g., housing, food) and *nonmortgage debt* (e.g., credit card bills).[6] I discuss nonmortgage debt in a later section. For now let's focus on the expenses we *need* to pay. The money left after doing that is called *discretionary income* (DI):

$$\text{DI} = \text{Disposable income} - \text{Necessary expenses}$$

$$= \text{Gross income} - \text{Income Taxes} - \text{Necessary expenses}. \quad (3.7)$$

Discretionary income is money you can spend at your discretion. You've already paid your rent (or mortgage), your electricity bill, and all other necessary expenses.

---

[5]Later I'll introduce you to a useful concept that keeps track of all your taxes. We'll use it to explain why it's easier to save by cutting expenses rather than increasing income.

[6]I'll talk about the other bills—like your cable and phone bills—later in the chapter.

If you keep track of your expenses (which I highly recommend you do), you may have noticed that your necessary expenses have been going up. There's a hidden economic force to blame; it drives up many, if not all, of our necessary expenses, and it happens so gradually that many of us don't notice. Worse yet, if you don't fight this force, it'll slowly erode your discretionary income. But there's hope. In the next two subsections I'll take you back in time to unmask this invisible economic villain and then show you how math can help us thwart its plans.

3.2.1   What a 19¢ Cheeseburger Can Teach You about Money

The year is 1955 and many iconic events are taking place that will shape the nation forever. In Montgomery, Alabama, a brave African American woman named Rosa Parks refuses to give up her bus seat for a white passenger, marking an important moment for the civil rights movement. In California, Mickey Mouse gets a permanent home at Disneyland. And in the small town of Des Plaines, Illinois, a man named Ray Kroc opens the first McDonald's restaurant.

The small menu at Kroc's restaurant offered hamburgers, cheeseburgers, fries, and beverages. The cost of the cheeseburger? 19¢ (before tax) [36]. Nineteen cents! Fast forward 60 years and today, in 2015, the McDonald's near your house sells cheeseburgers for $1 (before tax)—a 426% increase in price![*2] It's the same burger it was in 1955—one meat patty, cheese, buns, and condiments—so why is it 426% more expensive?

Before you call me a cheapskate, realize that it wasn't just McDonald's cheeseburgers that got more expensive since 1955; the prices of homes, cars, clothing, and most everything else has also gone up since then. One could describe this phenomenon as "a general increase in the overall price level of the goods and services in the economy" [37]. If one were an economist, that's what one would say—the quote is from the Federal Reserve's website, and they're describing *inflation*.

The Federal Reserve is the nation's central bank. It's charged with maintaining the stability of the financial system and making monetary policy decisions. "The Fed" does these things in part by influencing the supply of money in the economy through their ability to set the short-term interest rates banks and individuals pay for loans. (I'll tell you more about how that works in the next chapter.) For now, all we need to know is that the Fed shoots for a 2% yearly inflation rate; in their words, that inflation level is "most consistent over the longer run with the Federal Reserve's mandate for price stability and maximum employment" [38]. Here's what that means for us: there's a powerful entity (the Federal Reserve) whose job it is to make sure overall prices increase about 2% each year.[7]

The same Fed website is quick to mention that "inflation cannot be measured by an increase in the cost of one product or service." But let's do it anyway; let's see what the 426% price hike of the McDonald's cheeseburger can teach us.

To be fair, that 426% price hike took 60 years to happen. Most people would divide those two numbers and claim that the price increased by $426/60 \approx 7.1$ percent each year. *But that's not correct.* That's our first lesson. Let me show you why and then explain why it matters.

Let's assume I'm wrong and that the cheeseburger price did increase 7.1% every year since 1955. Then, a year later (in 1956) the price would've been 19¢ plus 7.1% of 19¢:

$$19¢ + (19¢)(0.071) = 19¢\,(1 + 0.071) = 19¢(1.071).$$

(I factored out the 19¢ in the second step.) Notice that I've converted 7.1% to decimal form by dividing by 100. Hereafter all calculations involving percentages will use the decimal form of those percentages. Next, notice that *next year's price is last year's price multiplied by* $1 +$ (decimal form of annual percentage increase). Using this rule the price another year later (in 1957) would be

$$19¢\,(1.071)\,(1.071) = 19¢\,(1.071)^2.$$

---

[7] "Why does the Fed insist on *inflation*? Why can't we have *deflation*, where prices go *down*?" Good question. Economists have no enemy greater than deflation, and here's why: would you buy *anything* today if you knew it'd be cheaper tomorrow? Neither would I. The ensuing decline in consumer spending would cause mass business closures and layoffs.

The "2" exponent on the right-hand side is doing double duty—it's tracking the number of years since 1955 and also tracking *how many times we've multiplied* 19¢ *by* 1.071. This observation helps us skip to the end of the calculation and say that the price 60 years later (in 2015) is 19¢ multiplied by 1.071 *60 times over*:

$$19¢\,(1.071)^{60} \approx \$11.64.$$

That's an expensive cheeseburger! The math is right, so that 7.1% annual price increase must be wrong. What we've learned is that when the price of something increases by X% each year for Y years the annual increase is not X/Y percent. So what's the correct answer?

Let's denote by x the unknown annual percentage increase. What we've learned then tells us that "19¢ increases to $1 (100¢) at x% annually for 60 years" is mathematically equivalent to the equation

$$19\,(1+x)^{60} = 100. \tag{3.8}$$

Solving for x gives $x \approx 2.8$ percent,[*3] much smaller than the 7.1% we got earlier. Yet if you pull up pictures of McDonald's menus over the years you'll notice that the cheeseburger price *doesn't* increase by 2.8% every year. But that's no longer a math error; it's a company changing prices as it sees fit. All (3.8) was designed to do was give us the constant yearly increase that grows 19¢ into 100¢ over 60 years. That's useful because (in our context) it gives us a rough idea of how inflation affected the cheeseburger price.[8]

Inflation also increases necessary expenses—like housing, food, and electricity. Unfortunately, these often rise in cost *faster* than the overall inflation rate. But if math can conquer taxes, it can conquer inflation, too. In the next section we use what we've learned to see how we can reduce inflation's fattening effect on our necessary expenses.

---

[8]The annual growth rate calculated via (3.8) is called the *compound annual growth rate (CAGR)* (also the *geometric mean*); it's used widely in investing (more on that in the next chapter).

3.2.2 Taming the Inflation Beast

Let me pick on one necessary expense that many of us have: rent. Even if you don't pay rent (e.g., if you have a mortgage) the stores you shop at pay rent to their landlords. Directly or indirectly, increases in rent make their way straight to your monthly budget. As anyone who's rented before knows, rents can increase a lot, and very quickly. Let me show you the data, and how math can help us eliminate yearly increases in rent.

Figure 3.2(a) shows the median monthly cost of rent plus utilities—called "gross rent" by the government—in the United States for each decade between 1940 and 2010.[9] The trend is clear: *up*, and more important, *not linear*. What type of nonlinear function is it? (Don't read the caption to the figure just yet!)

Let's start with what we know. Gross rent increased from $27 to $855 over the 70-year period between 1940 and 2010. Can you use what we did in the last section to calculate the yearly increase? I've put the details in the appendix;[*4] the answer is about 5%. That's *more than double* the Fed's 2% inflation target. If this trend continues into the future, what will the median gross rent be *t* years from now?

Here's how you figure that. First, pretend that houses are like cheeseburgers. Then, use the reasoning that led to (3.8). The result is that *t* years from now an apartment whose gross monthly rent is currently $1,000 would have a gross rent of

$$R = 1,000(1.05)^t. \tag{3.9}$$

This is an *exponential function*, so named because the independent variable (*t* in this case) appears as an exponent. When *t* is a whole number we can interpret $(1.05)^t$ just as we interpreted $(1.071)^{60}$ in the previous section: 1.05 multiplied by itself *t* times. (For example, $(1.05)^3 = (1.05)(1.05)(1.05)$.) This "go forth and multiply" aspect of (3.9) is part of what defines an exponential function. Indeed, exponential functions arise whenever a quantity is multiplied by itself again and again. Here's a more general definition.

---

[9]These are *median* prices, so some cities may have experienced higher (or lower) rents.

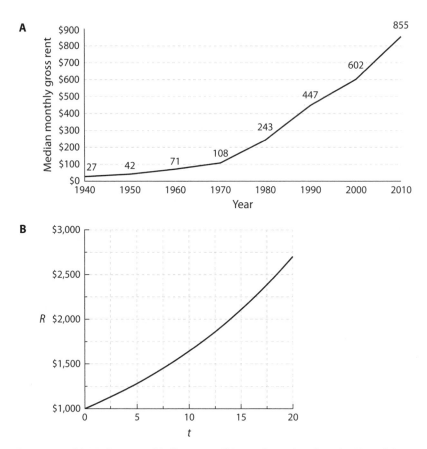

Figure 3.2. (a) Median monthly "gross rent" (rent plus utilities) in the United States between 1940 and 2010. Source: [39]. (b) A graph of the function $R = 1,000(1.05)^t$.

Introducing ... the Exponential Function

An equation of the form

$$y = ab^x$$

is called an *exponential function*. The number a is called the *initial value* (since when $x = 0$, $y = a$), the number b the *base* (we require $b > 0$ and $b \neq 1$), and the number x the *exponent*.

Based on this definition the initial value of the exponential function (3.9) is $a = 1,000$ and the base is $b = 1.05$. Since $b > 1$, we know that R increases as t increases (i.e., the monthly gross rent of our hypothetical apartment increases over time). Figure 3.2(b) shows the graph of the function; notice how closely it resembles the actual gross rent graph from part (a) of the figure.

We can also use this definition to extract the following interpretation for the base b: when the x-value increases by one unit, the y-value of an exponential function is multiplied by b.[*5] That y-values of exponential functions get multiplied by the base explains why exponential growth is much faster than linear growth, where the slope merely adds or subtracts from the y-value (see the appendix for an example).[*6] To appreciate how fast exponential growth can be, here's a fun bonus question. Which of the following options would you choose: $1 million dollars right now, or 30 days from now the result of doubling 1¢ every day for those 30 days? (Hint: Good things come to those who wait.[*7])

Okay, so inflation is driving up the cost of rent *exponentially*. It's likely having the same effect on the rest of your necessary expenses, too (though the yearly increases are different for each expense). That means your discretionary income (3.7) is at the risk of decreasing exponentially (yikes!) unless you can match the rising cost of your necessary expenses with at least as much of an increase in your after-tax income. But that's nearly impossible for most people (since it means making 2% to 5% more each year). So what ends up happening? Usually one of two things: someone else in the household gets a job or you take on debt. Senator Elizabeth Warren (a scholar of bankruptcy and commercial law) shows us that both have happened since 1970 in her book *The Two-Income Trap*. Warren gave a lecture describing the findings at the University of California, Berkeley, in 2007 (the lecture was videotaped; it's on YouTube under the title "The Coming Collapse of the Middle Class"). In it she presents data showing that necessary expenses since 1970 have gone up so much that they've forced households to have two income earners. Even more alarming, she points out that while in 1970 married couples required only 50% of one income to pay necessary expenses, today they require 75% of both incomes! That's precisely the dwindling discretionary income problem we just saw, and what Warren is telling us is that we've responded by sending another household member to work to keep up with rising necessary expenses.

But as powerful as inflation is, it's no match for math. Now that we understand the real cause of the problem—the exponential growth in necessary expenses—that itself suggests one strategy to tame the inflation beast: convert every inflation-affected expense into a fixed expense. For housing that means converting rent payments into mortgage payments (i.e., buying a home). If you choose a 30-year fixed-rate mortgage (the most popular choice) the monthly payments will *stay the same for 30 years*. After that you'll have paid off the loan and you'll stop paying for housing.[10] All we need to do now is figure out how much house you can buy so that the mortgage payment is equal to your current rent payment. If we can manage this, we'll remove inflation's ballooning effect on your housing expense. Let's consult the math.

The first thing we need is the formula for the monthly payment M on a loan of L dollars that charges a fixed annual interest rate of r% and requires n payments:[*8]

Calculating the Monthly Payment on a Fixed-Interest Rate Loan

$$M = \frac{Lc}{1 - (1+c)^{-n}}, \quad \text{where} \quad c = \frac{r}{12}. \qquad (3.10)$$

Here M is the monthly payment, L is the loan amount, r the annual interest rate (expressed in decimal form), and n is the number of monthly payments.

For a 30-year mortgage $n = 12(30) = 360$ (since you make 360 monthly payments). As an example, the monthly payment on a house bought for \$100,000 with a 30-year fixed-rate mortgage at 6% is about \$537.[*9]

Now, let's say your current monthly rent payment is P dollars. To estimate how much house you can afford we set $M = P$ in (3.10) and

---

[10]You'll still pay things like property tax, but we'll talk about that later.

TABLE 3.2.
Estimated home price (based on (3.11)) that one can afford by swapping a monthly rent
payment of P dollars for a 30-year fixed-rate mortgage of $r\%$. Note: see the appendix
for a discussion of how to account for other expenses, like property taxes.[10]

Current Monthly Rent	Interest Rate r (%)				
Payment P ($)	3	3.5	4	4.5	5
1,000	237,189	222,695	209,461	197,361	186,282
1,200	284,627	267,234	251,353	236,833	223,538
1,400	332,065	311,773	293,246	276,306	260,794
1,600	379,503	356,312	335,138	315,778	298,051
1,800	426,941	400,851	377,030	355,250	335,307
2,000	474,379	445,390	418,922	394,722	372,563

then solve for L. Doing so yields the formula

$$L = \frac{P\left(1 - (1+c)^{-n}\right)}{c}. \tag{3.11}$$

Table 3.2 shows a few combinations of P and r for a 30-year fixed-rate mortgage ($n = 360$), along with the associated L-value (the estimated home price) derived from (3.11). Notice that as the interest rate increases, the home prices one can afford decrease. But there's good news, especially for those of you paying high rents (e.g., the $2,000 monthly rents common in Boston and other big cities): trade your rent payment for a mortgage and you can buy up to a $475,000 house!

One quick note: owning a home requires paying property taxes and perhaps other expenses you don't incur as a renter. Incorporating that into our calculation lowers the values in Table 3.2; check the appendix to see how you can easily adjust the calculation.[10]

Converting a rent payment into a mortgage payment may also have another financial benefit: when you sell your house, you may end up making a profit! So, you may end up getting paid to live in your house! (That never happens when you rent; the most you get back is your security deposit.) Plus, if you lived in the home for at least 2 years you likely won't pay federal tax on the first $250,000 of profit from the sale if you file as single (or $500,000 if you file as married).

Buying a home is a big step, and the formulas I've presented are just one ingredient to consider.[11] But let me bring things back to how we got to discussing home ownership: you can combat inflation by converting your inflation-affected monthly expenses to fixed expenses. This may require creativity for your other expenses,[12] but the payoff is big— *perpetual* savings.

"But I can't buy a house right now," you might say. No problem; I've got another thing you can do to dramatically improve your finances: *eliminate your debts.*

3.2.3 How to Get Out of Debt Faster

As of February 2015, the total outstanding U.S. consumer debt—which includes credit card debt, student and car loans, but not mortgages— was $3.34 *trillion*. To get a feel for how mind-blowing that number is, consider this question: if you stacked $1 bills on top of each other, how high would a $3.34 trillion stack be? The answer: the stack *would almost reach the moon!*[13] Your nonmortgage debt stack is (hopefully) much shorter than that. But whatever debt you have, let me show you the fastest way to eliminate it, according to math.

First up, we need to know how long it takes to pay down debt. We're in luck here because (3.10) applies to any loan with monthly payments, be that a mortgage, a car loan, or a credit card debt (recall that the L in the formula represents the loan amount). In the previous section we solved for L to see how much house you could afford, but since we're now interested in how long it'll take to pay off the debt, we now want to solve for n—the total number of monthly payments. After a little algebra we get*[11]

$$(1 + c)^n = \frac{M}{M - Lc}. \tag{3.12}$$

---

[11] nytimes.com has a powerful calculator that compares the rent vs. own decision [40]. You should also consult a reputable realtor familiar with the current real estate market and the area you're thinking of buying into.

[12] For example, growing some of your own food and generating your own electricity (e.g., with solar or wind power).

[13] About 90% of the distance to the moon, to be more precise.

Notice that n is an *exponent*. So, to solve for n we need to "unexponentiate" the left-hand side. This is done with *logarithms*. Here's the definition of a logarithm.

The Definition of a Logarithm

An equation of the form

$$y = \log_b x,$$

where $b > 0$ and $b \neq 1$, is called a *logarithm*. The number b is called the *base* of the logarithm. When $b = 10$ we suppress the base and write $\log_{10} x$ as $\log x$.

The following fact describes how logarithms "unexponentiate" (see the appendix for a brief explanation):[*12]

$$\text{If} \quad y = b^x \quad \text{then} \quad x = \frac{\log y}{\log b}. \qquad (3.13)$$

For example if $y = 2^x$ then $x = \log y / \log 2$. This last equation says that the exponent, x, of the exponential function $y = 2^x$ is the ratio of the logarithm of y to the logarithm of 2.

Applying the equivalence (3.13) to (3.12) yields:[*13]

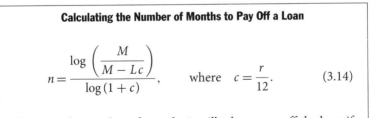

Calculating the Number of Months to Pay Off a Loan

$$n = \frac{\log\left(\dfrac{M}{M - Lc}\right)}{\log(1 + c)}, \qquad \text{where} \quad c = \frac{r}{12}. \qquad (3.14)$$

Here n is the number of months it will take to pay off the loan if M is the monthly payment, L the loan amount, and r the annual interest rate (expressed in decimal form).

For example, let's say you sign up for a new credit card and then go on a shopping spree. One month later your $1,000 bill arrives (so that $L = \$1,000$). The minimum monthly payment due is $M = \$20$, and the card's annual interest rate is 12% (so that $c = 0.12/12 = 0.01$). You don't have $1,000, so you decide to make the minimum payments. How long will it take you to pay off the card? Plugging in M and c into (3.14) gives us the answer: 70 months, or almost 6 years!

The story might be fictional, but the numbers I used are fairly realistic; credit cards often set their minimum monthly payment to be between 1% and 3% of the balance on the card—here the $20 payment is 2% of the $1,000 balance—and the average credit card interest rate these days is 15%. The takeaway: it takes a long time to pay off a debt if you just make the minimum payments.[14]

But (3.14) can also work in our favor. As M gets bigger (i.e., your monthly payments get larger) the payoff time *decreases*, and even small additions to the minimum payment can dramatically reduce the payoff time. To illustrate this, let's go back to our example and leave M as a variable in (3.14). Then the payoff time equation for your new credit card debt is

$$n = \frac{\log\left(\frac{M}{M - (1,000)(0.01)}\right)}{\log(1.01)} \approx 231 \log\left(\frac{M}{M - 10}\right). \quad (3.15)$$

Figure 3.3(a) shows a plot of this curve for $M \geq 20$.[15]

Notice that paying *just* $5 above the minimum each month (i.e., $25) reduces the payoff time by 18.4 months, about 1.5 years. Plus, you end up paying less in interest charges.

Now that we've worked out the math you can use it to conquer your own debts. You could, for example, adapt (3.15) to your debts and

[14]Besides, you'll also pay the most interest that way; to calculate the amount of interest you'd pay, subtract your total payments from the original balance. Continuing the example, by paying only the minimum each month you'd pay $20(70) − $1,000 = $400 in interest, or 40% of the original loan amount!

[15]Notice from the formula that M cannot equal 10 (we can't divide by zero, remember?). This is the math telling us that if we paid $10 a month we'd never pay off the card! We also can't have $M < 10$ because the logarithm of a negative number is not defined. So you need to pay more than $10 a month to the card to have any hope of paying it off.

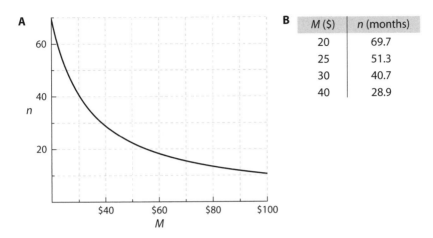

Figure 3.3. (a) The graph of the function (3.15). (b) A table of values of this function.

generate graphs like the one in Figure 3.3(a) to help you understand (and reduce) payoff times. (Check out the appendix for links to free websites that'll graph your payoff time equations.[*14]) That brings up a good question though: if you have multiple debts, which one should you focus on eliminating first? Math's got the answer: pay off the highest interest rate debt first and then pay off the one with the next highest rate, and so on. This approach minimizes the total interest paid.[*15]

One last strategy I'll share is the *debt snowball*: once you pay off one of your nonmortgage debts, you add those payments to the next debt; once you pay off that debt, apply its payments plus the original debt's payment to the next debt, and so on. Combine this with the "highest rate first" method and you've got a debt-destroying math-backed payment plan.

Once your debts are paid you'll have lots of uncommitted cash in your monthly budget. You could spend it, but a better option is to save it (or perhaps invest it). I'll delay the investing discussion to next chapter. In the next section I focus on the saving bit. I'll show you how increasing your yearly savings can help shorten your time to *financial independence* (FI)—that liberating date when you can quit your job and live off your savings alone. Yep, math can change your life!

3.3 HOW MANY YEARS WILL IT TAKE YOU TO REACH FINANCIAL INDEPENDENCE?

If you're going to quit your job, you'll need some other way to pay your *total expenses*, by which I mean

$$\text{Total expenses} = \text{Taxes} + \text{Necessary expenses}$$
$$+ \text{Nonmortgage debt payments}$$
$$+ \text{Discretionary expenses.} \qquad (3.16)$$

(The "discretionary expenses" term here includes expenses that aren't necessary and aren't debt payments; for example, your cable and phone bills fall into this category.) If you had a stash of cash sitting in an interest-earning account you'd be all set. But how much money would you need? And how much interest would the bank need to pay you? We'll answer these questions soon. But first let's talk about a little known fact: *reducing expenses, not boosting income, is the faster way to financial independence*. Here's a thought experiment to convince you of this.

Suppose you want to save $100. You have two options: cut your expenses by $100 or increase your income. If you increase your income you have to pay taxes on that new income, and that may include more than just federal income taxes. To quantify the total tax you'd pay, add up all the taxes you paid last year—federal, state, payroll, and so on—and then divide that amount by last year's total income. The percentage you get is called the *total effective tax rate*. Citizens for Tax Justice puts out a report each year titled "Who Pays Taxes in America?," and in the 2015 report they calculate that tax filers with an income of $48,900—roughly the average income in 2015—had a total effective tax rate of about 27% [34].[16] This means that 27% of these taxpayers' income was used to pay their taxes. If we assume your total effective tax rate is the 27% average rate, the extra $100 you earn would dwindle to $73 after taxes. That means you need to earn $100/0.73 ≈ $137 to be left with

---

[16]The Institute on Taxation and Economic Policy has all the data you could want on its website, even broken down state by state [35].

$100 after paying all taxes on that income. *That's 37% more than what you need to cut your expenses by ($100).*[17]

Hopefully I've now gotten you laser-focused on saving. Unfortunately, people in the United States are terrible savers. As of April 2015, the *personal savings rate*—the amount saved as a percentage of disposable income—was 5.6% (according to the Federal Reserve); in the 1970s that number hovered around 11%, almost twice what it is today.[18] That's bad news because your FI date is partly a function of how much of your income you save. Let me show you the math.

We first need to set up the assumptions (some of these may seem unrealistic, but I'll later indicate how to modify the calculation to make it more realistic):

- You save the same amount each year: S dollars per year.
- Your savings are invested and earn r% per year.
- You currently have B dollars saved that'll also be invested earning the same r% per year. (Note: B could be zero.)
- Your yearly expenses stay the same and are equal to your current total expenses (3.16) (same dollar amount, though the contributions of each category in (3.16) may change): T dollars per year.
- You have no other assets you can sell—no home, no car—or future sources of income (no retirement account, no Social Security, etc.).

This is all we need to calculate the FI date. The formula is complicated, so let me walk you through it in words first.

First, we calculate the value t years from now of your S dollars of yearly savings earning r% each year (this is called finding the *future value of an ordinary annuity*). We then calculate the balance t years from now of your B dollars of initial savings earning r% each year. Adding these totals yields the balance t years from now of your "nest egg"; let's call that balance N_t. A rule of thumb in retirement planning is to limit yearly withdrawals to 4% of your nest egg's value,[19] meaning

---

[17]The only way out is to not pay *any* tax on that $100.

[18]The savings rate drop may be due to inflation's effect on our necessary expenses, as Senator Warren's data implicitly suggest. Everything's connected!

[19]There's research to back this up, though taking into account inflation and stock market fluctuations may change this number. The *New York Times* recently discussed this in an article describing the history of the 4% rule and its alternatives [41].

you withdraw $0.04N_t$ each year to cover your yearly expenses of T dollars. All that remains is to set $0.04N_t$ equal to T and solve for t (see the appendix for the calculation).*[16] Doing so yields the number of years until financial independence:

Calculating the Number of Years until Financial Independence

$$t = \frac{\log\left(\dfrac{25r + STE}{STE + \frac{Br}{T}}\right)}{\log(1 + r)}. \tag{3.17}$$

Here t is the number of years until financial independence, $STE = S/T$ is the savings-to-expenses ratio (the ratio of your yearly savings S to your yearly total expenses T), B is your current savings balance, and r is the investing return rate (in decimal form).

This might be the scariest-looking equation we've seen thus far. But I promised one equation to rule them all, and this one, with its power to so dramatically change your life, certainly fits the bill. Let's discuss the many insights hidden in (3.17).

First, note that the *savings-to-expenses ratio, STE = S/T* in (3.17) is *not* the "personal savings rate" tracked by the Fed. Instead, it's just the ratio of yearly savings to yearly total expenses. Here are two hypotheticals to help you get a feel for the STE:

- If you save $1,000 this year but spend $10,000 then STE = $1,000/10,000 = 0.1$.
- If you save $8,000 this year but spend only $2,000 then STE = $8,000/2,000 = 4$.

Notice that if your STE value is less than 1 then you spent more than you saved, whereas if it's greater than 1 then you saved more than you spent.[20]

---

[20] STE can be expressed in terms of the percentage of your gross income you save (see item 17 in this chapter's appendix).

To see how important your yearly STE ratio is to lowering t—the number of years until financial independence—let's look at the $B = 0$ case of (3.17) (i.e., the case where you don't currently have any savings). Then (3.17) simplifies to

$$t = \frac{\log\left(\dfrac{25r}{\text{STE}} + 1\right)}{\log\left(1 + r\right)}. \tag{3.18}$$

Table 3.3 shows the t-values for various combinations of STE and r. Before discussing its two main insights, let me mention that I stopped at a return of 6% because that's roughly the long-term return, after inflation, that you can expect from investing in the stock market (we'll talk about why that's the case in the next chapter). On to the insights:

- **You'll likely be working forever if you spend 10 times (or more) what you save each year.** In these cases STE is at most 0.10, and according to the first row of the table, even with a 6% return it would still take you 48 years of saving to replace your expenses.
- **Even small improvements in your STE ratio can make a big difference.** Check out what happens when you go from an STE of 0.10 to 0.25—the FI date decreases *regardless of the return*, and it also occurs at least 15 years sooner!
- **Once you save at least what you spend each year, the rate of return doesn't matter much.** In these cases your STE is at least 1 (the last 5 rows of the table). Notice how the years until FI don't decrease too much as the return rate increases. Even at an STE of 1, the difference is 4 years between a measly return of 2% and a 6% return. Since low-return investments tend to be less risky than high-return ones, this suggests we focus more on saving than investing to move up our FI date.

These insights suggest that *the fastest way to reach financial independence is to increase your STE ratio.*

Table 3.3 is based on (3.18), which is a simplification of (3.17). If you already have savings, the numbers in Table 3.3 will be lower. Moreover, if the last assumption we made is not true—that you have no assets or other sources of savings—then that too will decrease your time to FI.

TABLE 3.3.
Years until financial independence depending on the yearly savings-to-expenses ratio and the annual return on invested savings. Note: values are rounded.

Savings to	Annual Return on Investment (%)				
Expenses Ratio	2	3	4	5	6
0.10	90	72	61	53	48
0.25	55	47	41	37	33
0.50	35	31	28	26	24
0.75	26	24	22	20	19
1	20	19	18	17	16
2	11.3	10.8	10.3	9.9	9.6
3	7.8	7.6	7.3	7.1	6.9
5	4.81	4.73	4.65	4.57	4.50
10	2.48	2.45	2.43	2.41	2.40

Finally, note that you can run the FI calculation at the end of each year and use whatever savings you've already accumulated as your B value. (This is one way to deal with the "constant yearly savings" assumption.) Tracking your progress to FI year to year may also motivate you to increase your STE ratio.

You know, every time I think about financial independence one side of me still thinks "yeah right, I'm gonna be working forever." But then I look back at (3.17) and I'm reminded once more that *mathematics can help us change our lives for the better*; it's that "math → empowerment" equation (1.9) at work again. The math in the first part of the book steered us toward specific foods that can make us healthier and live longer. Now, equation (3.17) is giving us a way to reach a new stage in our lives, defined not by where and how long we work but by the freedom to do the things that matter most to us without worrying about earning an income. I think it's amazing that math can help with all that.

But first, according to (3.17), you have to increase your STE ratio. That's where the rest of our work in this chapter comes in. It's my hope that the tax-reducing, inflation-taming, and debt-eliminating strategies math has revealed apply to your situation, and that they help you boost your STE ratio and shorten your time to financial independence. I've assumed that you'll be investing your accumulated savings and I owe you a discussion of why 6% is perhaps the most reasonable long-term return you can expect. I'll also sweeten the deal and describe multiple investment strategies that drastically reduce the risk of loss and help limit losses when they do occur. So, for our last stop on this journey through personal finances, let's head into the world of investing.

Chapter 3 Summary

MATHEMATICAL TAKEAWAYS

- Piecewise linear functions are functions made up of pieces of lines.
- Exponential functions arise when a quantity is growing (or decaying) by x% each period (e.g., year, day). The graphs of exponential growth functions curve upward, and their y-values increase much faster than those of linearly growing functions.
- Logarithmic functions show up when one tries to solve for the exponent in equations involving exponential functions (see (3.13)).
- Logarithmic functions, like exponential functions, are defined by their base; the base 10 logarithm is one of the most common ones and is denoted by $\log x$.

NONMATHEMATICAL TAKEAWAYS

- Federal tax due depends on your taxable income, your filing status, and your tax bracket. Your tax bracket determines how much of each dollar of additional taxable income you get to keep (see Figure 3.1(b)).
- Increasing deductions decreases your federal tax due according to your tax bracket, whereas tax credits decrease your federal tax due "dollar for dollar."
- Your total effective tax rate—calculated as the ratio of total taxes paid to total income—determines how much of your total income you get to keep.
- Many of your necessary expenses may go up over time due to inflation, the overall rise in prices of the goods and services in the economy. You can fight inflation by converting your inflation-affected expenses into fixed expenses (for example, rent into a mortgage).
- Even small increases in the monthly payment to a debt can shave off months, if not years, from the payoff date.

- If you have multiple debts with different interest rates, paying off the one with the highest rate first will minimize the total interest you pay.
- As you pay off debts use the freed-up cash to increase payments to other debts (the "debt snowball" method).
- Cutting your expenses is a better way to save than earning more income; income is subject to income taxes, which, depending on your total effective tax rate, can take away anywhere from 25% to 40% of the additional income earned.
- Increasing your yearly savings-to-expenses ratio (STE) will get you to financial independence sooner.
- As your STE ratio rises, the return rate on your invested savings matters less and less.

BONUS: A FEW PRACTICAL TIPS

- *Track your spending.* The first step to getting your finances under control is to figure out where your money is going. You can do this with pencil and paper, a spreadsheet, apps, or budgeting websites.
- *Get rewards credit cards.* This is an easy way to make extra money every year. Many cards give you 1% to 2% cash back on virtually all of your purchases. Some even give you up to 5% cash back on purchases in certain categories (e.g., groceries). This means that for every $1,000 you spend you'll get between $10 and $50 back.
- *Get rewards checking/savings accounts.* Many credit unions offer up to 3% annual interest for opening a checking account and making a minimum number of purchases each month with your debit card; some banks also run similar promotions.

■■■■■■■■■■■■■■■■

CHAPTER 4

How to Beat Wall Street at Its Own Game

MY FIRST INTRODUCTION TO INVESTING WAS IN COLLEGE. I had a friend named Nathan who lived down the hall from me. One day, while walking by Nathan's room, I saw his computer screen through the open door. I remember seeing flashing green and red boxes everywhere and asking Nathan what was going on. Turns out he was watching his investments as they fluctuated in value—a green flash meant he'd just made money, a red one that he'd lost money. Nathan mentioned he was up over $1,000 on the week, and he told me how easy it was to start investing. I took the plunge and opened an investment account shortly afterward. I had no idea what to invest in. So I bought stock in Intel— the company that makes the CPUs that power many computers. Within days I was *up* $150 (yeah!). Then, a few weeks after that, I was *down* $150 (ugh!). "Should I wait for it to go back up or just cut my losses?" I kept thinking. The days passed and my account value kept dropping. When my losses passed the $200 mark I finally threw in the towel. One word: *fail*.

Nathan hadn't told me you could *lose* money investing. I should've known that, but I got caught up in the "easy money" euphoria. Sure enough, after actually reading through the documents I signed to open my account, I found one phrase inserted everywhere: "Investing involves risk, including possible loss of principal."

I've learned a lot since those days (including that you should read *every* word of any document you sign before you sign it). For one, investing has a lot in common with gambling. In both cases you show up with your hard-earned money, hoping to make a profit. But unless you're familiar with the game you're playing—how it works, what the odds of success are—you'll be at the mercy of luck alone. Fortunately, investing is another thing math can help with.

In this chapter I describe the various insights that math can offer us regarding investing. I'll introduce you to various types of investments across the risk spectrum and describe how math helps quantify their risk. We'll discuss how to combine these investments into a portfolio with a prescribed risk level. After using some more math, I'll present you with a simple portfolio with an average yearly return of 10.2% since 1987, and which had positive returns over *any* 5-year (or greater) holding period since then! Then, I'll tell you about a portfolio that had average annual returns after inflation of about 6% since 1926, *regardless of whether the economy was in recession or expansion*!

4.1 HOW TO MAKE 15% A YEAR, *GUARANTEED*

Let me start off by offering you a way to make up to 15% a year with *zero* risk: pay off your nonmortgage debts.

I know, this isn't the hot stock tip you were expecting. But think about it. The least risky investments return a measly 1% to 3% a year (we'll discuss this in the next section). Plus, many of those investments don't let you access your funds until years later unless you pay an "early withdrawal" penalty (which then reduces the return). And don't forget taxes; your investment earnings add to your income and increase your tax burden, effectively lowering the return.

Paying off debt, on the other hand, helps you avoid paying interest. The interest you would've paid is therefore the return you get from this "investment." The more interest you save, the greater your return on investment (ROI). Plus, there are no taxes on your "earnings," you're guaranteed to make money, and since you decide how much to pay each month you have greater control over the funds you're "investing." So before you invest, pay off those debts!

There is *one* scenario where it makes more sense to invest rather than pay off debt. The rough rule of thumb here is that investing money you would've used to pay down debt only makes sense when

Investment return ≥ 1.2(Debt's annual interest rate).

(See the appendix for how I came up with this.*[1]) For example, suppose you're carrying debt on a credit card with an annual interest rate of 15%. Unless you can find a safe investment with a return of at least 18% (1.2 times 15), math says you're better off using all money to pay off the debt. But even if my rule of thumb does apply, here are two more great reasons to use that investing money to pay off your debts first.

- **Paying down debt is easier than investing.** It's risk-free, its returns are tax-free, and you don't need an investment account to do it.
- **One way to beat Wall Street is to not play its game.** Remember the warining about the possible loss of principal? Well you don't have to worry about losing your initial investment when you pay down debt— each payment lowers your balance; you're always making money!

Now that we've discussed a *risk-free* form of investing let's climb the risk ladder and discuss more traditional methods of investing.

4.2 THE SAFEST INVESTMENTS

The most traditional way to earn a return is to open a savings account at a bank or credit union (a member-owned version of a bank). It's a safe way to invest because almost all banks and credit unions carry deposit insurance—FDIC for banks and SIPC for credit unions. (If your bank/credit union doesn't have these protections, get out of there!) This means that anywhere from $100,000 to $250,000 of your checking and savings account balances are insured against your bank/credit union failing.[1] But this "principal protection" comes at a price: low returns.

Deposit institutions tend to set a savings account's return rate based on the Federal Reserve's *discount rate*, the "interest rate charged to commercial banks and other depository institutions on loans they receive from their regional Federal Reserve Bank's lending facility" [42]. Translation: the Fed loans banks money at the discount rate. As of this writing that's 0.75%. So if *you* loan the bank money—by opening a savings account, a CD, or any other similar "traditional savings vehicle"—it wouldn't make sense for them to pay you more than 0.75%

---

[1] The deposit insurance is per account.

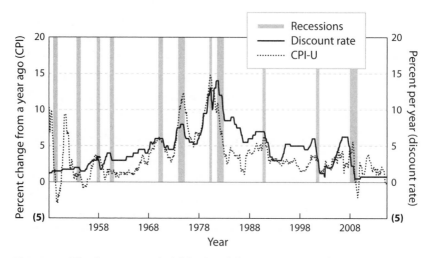

Figure 4.1. The discount rate (solid line) and the 1-year percent change in the Consumer Price Index for all urban Consumers (CPI-U) since 1948. Economic recessions are marked by the shaded regions. Data source: Federal Reserve Bank of St. Louis.

(since they can always pay the Fed 0.75% to borrow their money). This explains why the current average rate paid by deposit institutions for traditional savings vehicles ranges from 0.06% to 0.79% [43].[2]

Savings rates weren't always so low. Figure 4.1 gives us a little perspective. You could get up to 6% returns on your savings as recently as 2007 (as indicated by the solid line in the graph). Then the Great Recession of 2008–2009 came along (the gray regions in Figure 4.1 indicate recessions) and the discount rate plummeted to about the 0.75% it's been at since 2010. In fact, there seems to be a general trend when you look at all of the recessions: the discount rate goes up *before* the recession and *down* during or after. If we rephrase this in terms of what the discount rate means we discover something noteworthy about our economy: during and after a recession, the Federal Reserve makes it easier for banks to borrow money from them. When the recession ends and the economy starts expanding again, the Fed makes it more costly for banks to borrow money from them. (We alluded to this in

---

[2]To get the highest rate you need to invest in a 5-year *certificate of deposit (CD)*. In doing so you loan the bank money for 5 years, at which point they give it back to you with interest. (You can get your money out sooner if you pay the early withdrawal penalty.)

Section 3.2.1 when we mentioned that the Fed influences the supply of money in the economy; now you know one mechanism it uses.)

You might have noticed the 14% peak in the discount rate in Figure 4.1 and thought "a great return *and* principal protection? Sign me up!" Not so fast. A high return on your savings won't do you any good in periods of high inflation. That's why it's better to look at the *real return*—the return minus the inflation rate. Economists like to use the *Consumer Price Index (CPI)* as a gauge of inflation,[3] and as the dashed line in Figure 4.1 shows, the inflation rate is usually just below the discount rate. Translation: the highest real return of traditional savings vehicles is never more than a few percentage points above the inflation rate. And sometimes the real return is negative; at those times people lose money by saving using traditional savings vehicles! (They still earn interest on their savings, but not enough to offset the increases in the prices of the "overall goods and services in the economy.") The low—and at times negative—real returns of traditional savings vehicles led Warren Buffett (the third wealthiest person in the world as of this writing) to describe them as "the most dangerous of assets," pointing out that "over the past century these instruments have destroyed the purchasing power of investors in many countries" [45]. The takeaway: don't expect too high a real return (or even a positive one!) from traditional savings vehicles (e.g., a savings account, a CD). That's a bummer, especially considering that these are risk-free investments.[4] If we want *real* real returns then, we need to climb further up the risk ladder. We'll lose the principal protection, but as we'll see in the next section, math can help us control the risk of losing money.

4.3 QUANTIFYING INVESTMENT RISK AND RETURN

Every year in Smallville, two small businesses compete for bragging rights. Chloe, a soccer-playing 9-year-old, sells coffee on Returns Street.

---

[3] It's calculated based on the result of large surveys of families and individuals regarding their total expenses [44].

[4] There's always the danger that the FDIC/SIPC may become insolvent, so there is technically some risk.

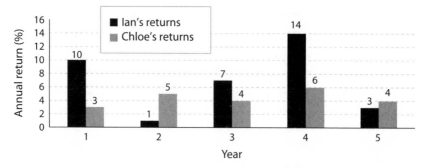

Figure 4.2. Annual returns over the past 5 years for Chloe's Coffee and Ian's Ice Cream; the numbers above the bars are the annual returns for that year.

Ian, a 10-year-old who collects stamps, sells ice cream on the other side of the street, directly opposite from Chloe. Figure 4.2 shows these young entrepreneurs' annual returns over the past 5 years.[5]

Chloe's returns look consistent (people always want coffee). But Ian's revenues depend on how hot the summer is, so his returns are more volatile. For now let's take this as our working definition of investment risk: *volatility of annual returns*. (After all, wouldn't you consider an investment that returns the same amount year in and year out low risk?) Given what you know, whose business would you invest in?

Well, the "eyeball test" we applied to Figure 4.2 is too qualitative. We need a *quantitative* measure of volatility to better evaluate the risk of investing in either business. Part of what makes Chloe's returns seem less volatile is that they're about the same every year—roughly 4% on average. Plus, the returns in years 1, 2, and 4 don't deviate too much from the 4% average.

Aha! The *deviation from the average return* seems to be a good metric for volatility. Statisticians call this concept the *standard deviation* of a data set. Here's how to calculate the standard deviation of the returns in Figure 4.2:

1. First, find the average. In Ian's case that's 7%, and for Chloe's it's 4.4%.

---

[5]Chloe and Ian calculate their returns by dividing the total revenue over the year by the total costs incurred that year, then subtracting 1.

2. Next, subtract the average from each data point and square the result. Then add together these quantities. For Chloe's returns, this yields

$$(3 - 4.4)^2 + (5 - 4.4)^2 + (4 - 4.4)^2 + (6 - 4.4)^2 + (4 - 4.4)^2 = 5.2.$$

3. Finally, divide your result by the number of data points minus one, and then take the square root. In Chloe's case we divide by 4 (5 data points minus 1) and then take the square root to get 1.14.

We denote standard deviation by the Greek letter sigma: σ. We just calculated that $\sigma_{Chloe} = 1.14$, and a similar calculation yields $\sigma_{Ian} = 5.24$. So, Ian's returns are indeed more volatile than Chloe's (about 4.6 times more volatile according to the standard deviations). Rephrased in terms of the calculation itself, we can say that Ian's annual returns deviated from their average much more than Chloe's. Score: Chloe 1, Ian 0.

"But Ian had a higher average return, so he probably made more money over the 5 years" you might say, and you'd be right. Here's how we calculate the *total return* over the 5-year period:

1. Convert each return to decimal form, add one, and multiply them. For Chloe's business this yields

$$(1.03)(1.05)(1.04)(1.06)(1.04) \approx 1.24.$$

2. Finally, subtract 1 and multiply by 100. The result is the total return over the investment period.

Chloe's total return was 24%, but Ian managed about a 40% total return, indeed higher than Chloe's. Score: Chloe 1, Ian 1.

We can now measure both investment risk (using standard deviation) and reward (using either the average return or the total return).[6] In Ian's case we found higher risk ($\sigma_{Ian} = 5.24$) and higher reward (average return of 7%), and in Chloe's case we found lower risk ($\sigma_{Chloe} = 1.14$) and lower reward (average return of 4.4%). This risk-reward trade-off is a common feature of investing—if you're willing to

---

[6]There are other measures of risk and reward, but these are the most commonly cited.

TABLE 4.1.
Average return, standard deviation, and risk-reward ratio for the returns in Figure 4.2.

Characteristic	Investment	
	Ian	Chloe
Average return (%)	7	4.4
σ	5.24	1.14
RR ratio	1.36	3.86

take on more risk you'll likely get a higher return. But how much more reward can you expect from taking on more risk?

One concept that can help is the *risk-reward ratio* (RR ratio):

$$RR = \frac{\text{Average investment return}}{\text{Investment standard deviation}}. \tag{4.1}$$

(I hope this reminds you of the concept of energy density (Section 2.2); it's the nutrition-money analogy at work once again.) Table 4.1 shows the RR ratio for Ian and Chloe's returns, along with the other numbers we've calculated.

The 1.36 RR ratio of Ian's business means that each unit of risk resulted in just 1.36% annual return. Contrast that to Chloe's case, where each unit of risk resulted in 3.86% annual return. In other words: for the same amount of risk Chloe's business returned more than Ian's. That makes investing in Chloe's business a better "bang for your buck" decision. "But her returns are almost *half* that of Ian's!" I know, but Ian's returns are also more volatile. Score: still tied because there's no clear winner this time.

Here's a good idea: since each business has its merits, let's invest in both! Let's use math to do *two* things at once: build a portfolio that invests in both businesses and controls the volatility. Here's how we can do that.

Let's imagine I invest x% of my money in Ian's business and $(100 - x)$% in Chloe's business. I'm also going to *rebalance* my portfolio every year, meaning that at the end of the year I buy or sell accordingly to get back to an x% allocation to Ian's business. Rebalancing is important for two reasons: (1) if you don't rebalance portfolios, then the higher risk component may become a larger and larger share of the portfolio over time, and (2) rebalancing ensures that you implement the most basic rule of investing: buy low and sell high.[7] Here are the risk/reward stats for these "Ian & Chloe" portfolios:

$$\text{Average Return} = 0.026x + 4.4,$$

$$\sigma = 0.043x + 0.78. \qquad (4.2)$$

Linear functions, again! But in this case they're linear because I made them so—I used a spreadsheet to run various scenarios where x ranged from 0% to 100%, and then fitted linear functions to the data (this explains, for example, why when $x = 0$ or $x = 100$ we don't get the exact standard deviations in Table 4.1). Trust me, though, the equations in (4.2) are very good approximations to the risk/reward characteristics of the combined portfolio for x-values between 0 and 100.

Since the equations in (4.2) are linear functions, we can apply our slope interpretation from Chapter 1 to conclude that for each 1% increase in the portfolio's allocation to Ian's business the average return increases by 0.026% and the volatility increases by 0.043. (Since the increase in volatility is higher than the increase in return, each increase in x results in a decrease of the risk-reward ratio.) Thus, (4.2) allows us to build a portfolio with a prescribed risk level by setting σ equal to our maximum volatility threshold and solving for x. In the next section we'll use this "mix-and-match" approach to build the portfolios I alluded to at the start of the chapter.

4.4 STOCKS, BONDS, AND THE "ALL-WEATHER" PORTFOLIO

The portfolios I'll describe invest in stocks and bonds.[8] More specifically, they invest in *exchange-traded funds* (ETFs). An ETF generally invests in multiple securities at once (like a mutual fund); this gives

---

[7] See [46] for a discussion of other rebalancing strategies.
[8] You're probably already familiar with stocks, but perhaps not bonds. I describe what these are (as well as other terms I'll be using shortly) in item 2 of this chapter's appendix.

you instant *diversification*, which helps insulate your portfolio in the event that one particular stock the ETF holds has a bad day.[9] Also, ETFs can be traded any time the stock market is open; unlike a mutual fund (which only trades at the close of the market), this gives you the flexibility to sell your shares whenever you like.

With those preliminaries out of the way, here are the two components of this "all-weather" portfolio (a portfolio that does well in recessions *and* expansions):

- **An S&P 500 ETF.** The S&P 500 is a stock market index composed of 500 of the largest publicly traded companies in the United States. Two popular ETFs that reproduce the returns of the S&P 500 index nearly identically are State Street's SPY and Vanguard's VOO.
- **A long-term government bond ETF.** Two popular options are Vanguard's VGLT and the iShares 20+ Year Treasury Bond ETF (TLT); these track indexes of U.S. Treasury bonds maturing at least 10 (respectively, 20) years from now.

We get instant diversification in stocks with the S&P 500 ETF and low default risk in bonds from the Treasury bond ETF. Moreover, since bond prices tend to go up when stock prices go down, the bond component of the portfolio should offset any stock losses in down years (e.g., 2008). Finally, I will also assume that we rebalance the portfolio annually to an $x\%$ allocation to stocks (like we did in the previous section). How would these simple portfolios have done since 1987 (I'll explain why I chose that date shortly)? Spectacularly well. Table 4.2 shows the details; here are the highlights:

- The first line in the table shows the results of investing in *only* long-term Treasury bond funds (since the stock allocation is 0%). An almost 9% average annual return is pretty good, as is a maximum 1-year loss of 13.4% over *any* 1-year holding period starting between 1/1/1987 and 6/15/2014. But check out the RR ratio; it's *lower* than some others

---

[9]There are still other risks to investing in ETFs. One big one: ETFs are issued by companies, so there's always the risk of the ETF issuer going bust. Every ETF has a *prospectus* that outlines the overall risks of investing in that ETF; it's required reading, in my opinion, before you give someone else your hard-earned cash.

TABLE 4.2.

Portfolio stats assuming an x% allocation to an S&P 500 ETF, the remaining allocation to a long-term Treasury bond ETF, and annual rebalancing. The last column shows the maximum loss over any 1-year period starting between January 1, 1987, and June 15, 2014. All calculations are based on Yahoo! Finance data for VFINX and VUSTX, two Vanguard funds that closely track the performance of the S&P 500 and long-term treasuries, respectively.

Percent Allocation to S&P 500 ETF	Portfolio Stats			
	Average Annual Return (%)	σ	RR Ratio	Max 1-Year Loss (%)
0	8.8	11.6	0.75	13.4
10	9.1	10.3	0.88	10.3
20	9.3	9.3	1	8.1
30	9.6	8.8	1.09	8.4
40	9.9	8.9	1.11	12.5
50	10.2	9.6	1.06	17.5
60	10.5	10.7	0.98	22.6
70	10.8	12.2	0.89	27.8
80	11.1	13.9	0.80	32.9
90	11.4	15.7	0.73	38.1
100	11.7	17.7	0.66	43.3

in the table. Takeaway: you'd have gotten more reward for each unit of risk by *adding* stocks to the portfolio.

- But don't take that advice to mean that the 100% stock portfolio is the best one; look at what happens in the last line of the table. Sure, you got a much higher average annual return (11.7%), but with much more volatility and a maximum 1-year loss of 43.3% (ouch!). In addition, that portfolio's RR ratio (0.66) was lower than the all-bond portfolio.

- Now, go back and scan the RR Ratio column from top to bottom. You'll notice that the RR ratio increases up until about $x = 40$%.[10] This is a classic 60/40 bond/stock portfolio. Increasing the allocation to stocks beyond 40% still increases return, but increases volatility more. Plus, the maximum 1-year losses start getting unbearable.

- Here are the linear (and quadratic!) functions of x that describe each portfolio's risk and reward (these are the linear and quadratic

---

[10]I rounded those numbers; the maximum RR ratio occurs at $x = 38$%.

functions that best fit the data in Table 4.2):

$$\text{Average return} = 0.023x + 8.76,$$

$$\sigma = 0.002x^2 - 0.14x + 11.35,$$

$$\text{Max 1-year loss} = 0.012x^2 - 0.51x + 13.65. \qquad (4.3)$$

These functions fit the table's data well, but in full disclosure I only used the table data up to $x = 50$ for the last one (the new "Max 1-year loss" function), since I wouldn't want to lose more than 20% of my portfolio in any given year. There are two noteworthy things about this new function: (1) it allows us to determine the stock allocation (x) based on the maximum 1-year loss we're willing to endure, and (2) since it's a quadratic function, in general you'll get two x-values for a given maximum 1-year loss value, allowing you to choose the one with the highest return while capping your maximum 1-year loss. For example, if you don't want to lose more than 10% in one year, we'd set the max 1-year loss function equal to 10 and solve for x. We get $x \approx 9.1$ and $x \approx 33.3$. Looking back at Table 4.2, the portfolio with $x = 10$ has an average annual return of 9.1%, and the one with $x = 30$ has a return of 9.6%. So the 30% stock portfolio would have capped 1-year losses to roughly 10% but with a higher return than the 10% stock portfolio. Pretty neat!

• Let me single out the 50/50 portfolio for a moment. Figure 4.3(a) compares its annual returns (shown in black) to that of the S&P 500 and long-term Treasury bonds. As you can see, it did a great job of capturing the upside returns with little downside losses. Figure 4.3(b) shows the compound annual growth rate (CAGR; recall the definition in Section 3.2.1) for different holding periods. Check out the black line first; of *all* 3-year holding periods since 1987 the portfolio's return was only negative about 3.6% of the time (between 2002 and 2003 it briefly turned negative, and again shortly after 2008). As you extend the holding period the results get better—returns were positive over every 5-year period since 1987 (the gray line), and over every 10-year period they were positive and less volatile (the dotted line). This illustrates another investing rule of thumb: *if your portfolio is diversified enough, the longer your holding period the smaller the volatility in your returns.*

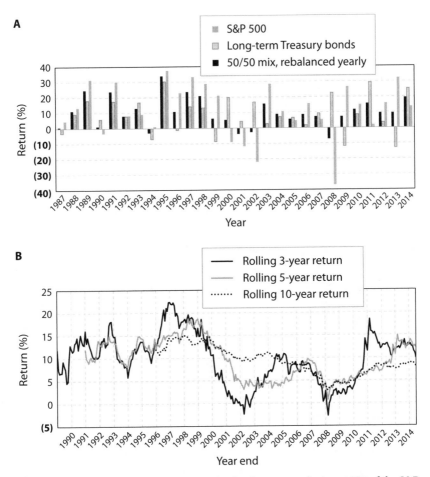

Figure 4.3. (a) The annual total returns (dividends reinvested) since 1987 of the S&P 500 (gray bars), long-term Treasury bonds (textured gray bars), and a portfolio invested in each equally and rebalanced yearly (black bars). (b) The 50/50 portfolio's compound annual growth rate (CAGR) for different holding periods: 3 years (black), 5 years (gray), and 10 years (dotted).

- Finally, I'd like to emphasize that the 50/50 portfolio did well despite the many challenges the economy faced since 1987. These included the 22% one-day drop in the stock market on October 19, 1987, the dot-com bubble burst (and ensuing recession) of 2000–2002, and the Great Recession of 2008.

There you have it; I made good on the promise of delivering a historically high-return (10.2%), low-volatility portfolio—the 50/50 stock/long-term government bond portfolio. But at the end of the previous chapter I mentioned that 6% (not 10.2%) is a more reasonable long-term return. Why the difference in numbers?

The answer has to do with one major "but" you may have had while reading the analysis above: "But you only gave me data since 1987." That's because that was all the data I could get for free. Other people have access to more data. The folks at Vanguard funds, in particular, recently did a study that found something amazing [47]: "Since 1926, the average annual returns of a portfolio invested in 50% stocks and 50% bonds was 7.75% during recessions and 9.9% during expansions." They also found that inflation reduced those returns to 5.26% and 5.59%, respectively.[11] In their own words: "the average real returns of such a portfolio since 1926 have been *statistically equivalent regardless of whether the U.S. economy was in or out of recession*" (emphasis original). Translation: since 1926 we had a Great Depression, a world war, quite a few other wars, an inflation crisis in the 1970s, major stock market crashes in 2000 and 2008, and yet this 50/50 portfolio delivered virtually the same average inflation-adjusted return through good years and bad—roughly 6%. That, ladies and gentleman, is a true "all-weather" portfolio.

In this chapter we've seen how math can help us deal with the inherent unpredictability of the stock market and the economy. Sure, investing involves risk; sure, "past performance is not a guarantee of future returns," as every fund prospectus states; but as we've learned, risk and return can be quantified. Once you do that, the math we've discussed can help you compare investments, combine them to create portfolios, and understand the historical risk/reward characteristics of those portfolios. In addition, you could follow my lead with (4.3) and create portfolios based on your tolerance for volatility. This all reminds me of the MIT Blackjack Team. This group of students used math

---

[11] The Vanguard study's bond component is more diversified (and lower returning) than the long-term Treasury bond component of the 50/50 portfolio in Figure 4.3. One way to replicate it using ETFs is to replace the TLT component of our 50/50 portfolio with AGG, the iShares Core U.S. Aggregate Bond ETF; it tracks an index of U.S. investment-grade bonds (which includes corporate and government bonds) of varying maturities.

to beat casinos at blackjack throughout the 1980s and '90s. Likewise, despite being just individual investors, we too can use mathematics profitably. We may not reap the multimillion-dollar profits that some on Wall Street do, but in our own microcosm, our efforts may shave a year off our financial independence date and give us that much more time to spend with those we love.

Speaking of those we love, it would be nice if that included not just our family but someone we're *in love* with. Love is another aspect of life that involves risk and volatility. And once again, math has much to say. But standard deviation and annual returns ain't gonna cut it—we'll need new mathematics to study love. In the last part of the book I'll draw on the mathematics of probability, dynamical systems, and game theory to help shed light on the problem of finding—and keeping—your soulmate. We'll discover practical strategies that can keep your relationship healthy over the long term, and also develop a systematic, fair, and equitable way to make joint decisions. Last stop: the mathematics of love.

Chapter 4 Summary

MATHEMATICAL TAKEAWAYS

- The standard deviation of a set of numbers is a measure of the volatility of that data set with respect to the average of those numbers.
- The standard deviation and annual returns of a two-asset portfolio can be expressed as functions of the percentage allocation to one of the assets. This makes it possible to adjust the allocation of the two assets to maximize return, minimize risk, or maximize the risk-reward ratio.

NONMATHEMATICAL TAKEAWAYS

- Above all else, remember these two important rules of investing: "investing involves risk, including the loss of principal," and "past performance is not a guarantee of future returns."
- The only truly risk-less "investment" that also happens to (potentially) provide a high, tax-free return is paying off your debts. There are,

however, rare cases when it might make sense to invest rather than pay off debt (see Section 4.1).

- Savings vehicles—savings accounts, bank CDs—generally carry deposit insurance, meaning that you'll get up to a predetermined amount of your initial investment back in the event the financial institution goes bankrupt. But this peace of mind comes with a cost: low returns. Sometimes these returns are lower than the inflation rate at the time, leading to *negative* real returns.

- There's a close correlation between the return on savings vehicles and the Fed's discount rate—the two tend to rise and fall together; see Section 4.2 for an explanation.

- Standard deviation helps quantify the volatility of an investment's annual returns; the larger the standard deviation, the larger the volatility.

- The risk-reward ratio—the ratio of an investment's average annual return to its standard deviation—is useful for comparing different investments. Those with high RR ratios have greater returns for the same volatility than those with lower RR ratios.

- Rebalancing portfolios is important, because otherwise the riskiest component of the portfolio may, over time, grow to constitute the majority of the portfolio.

- Investing in individual stocks or bonds is very risky. A better option is to diversify by investing in multiple stocks or bonds at once; ETFs— exchange-traded funds—are a practical way to achieve diversification.

- Portfolios that combine stocks with bonds sometimes have better RR ratios than all-stock and all-bond portfolios.

- It's possible to build portfolios with a given level of volatility or a given maximum 1-year loss.

- A portfolio consisting of 50% diversified stock and 50% diversified bond holdings has historically provided average real returns of about 5.5% regardless of the economic climate.

BONUS: A FEW PRACTICAL TIPS

- *Read every document.* Financial documents are a pain to read, I know. But please do read them before you sign anything.

- *Listen to podcasts to learn more about personal finance.* Podcasts are a great (and free!) way to learn more about your finances and the broader economy. One of my personal favorites is National Public Radio's *Planet Money* podcast.
- *Trade paper money first before you start investing.* Spend a few months (maybe even a year) "paper trading," where you use a spreadsheet to keep track of whatever investing strategy you've decided on without actually investing yet. Track your profits, losses, and volatility and try to take it seriously; it might help you discover what your real risk tolerance is.
- *Shop around for a brokerage firm, when you're ready to start investing.* Some charge as little as $5 for each trade, but these brokerages may not have the best customer service. Others have all the bells and whistles, but charge higher commissions. Think about how you'll use the brokerage—will you just use the online platform to execute your trades, or do you want the option to discuss trading strategies with someone at the firm?
- *Don't invest money you think you'll need sometime soon.* "Soon" is hard to quantify. But it's probably not a good idea to invest your emergency fund in anything riskier than a savings account, for example.

PART III

Looking for Love? There May Be an Equation for That

CHAPTER 5

■■■■■■■■■■■■■■■■

Finding "The 1"

MATHEMATICS AND SCIENCE ARE FOUNDED on deductive reasoning and the scientific method, respectively. But love, being an emotion, is rarely logical. This makes mathematizing love especially tricky. For that reason I'll be more explicit about the limitations of the mathematical models and results we'll soon discuss.

That was my disclaimer. Now on to the fun stuff.

In this part of the book I'll introduce you to a few mathematical models of dating and relationships. The models discussed in this chapter will focus on *finding* "the one"—that is, the actual search process. In the next chapter we'll talk about *keeping* "the one"—building a lasting, happy relationship. I'll kick things off by illustrating an entirely rational approach to finding a partner that, as with many love stories, may work fantastically well or be doomed to fail from the onset. All that uncertainty will lead us into the mathematics of randomness: *probability*. This newcomer will help us derive two powerful mathematical results: a step-by-step procedure for deciding when to stop dating and "settle down," and an algorithm for pairing people that produces couples who *will never cheat on each other*. So let's see what this atypical love doctor (math) has to say.

5.1 WHAT THE SEARCH FOR ALIENS CAN TEACH YOU ABOUT FINDING YOUR SOULMATE

"What type of person are you looking for?" That's probably a question your best friend has asked you before (or that you've asked yourself). Everyone has a different way of answering it—pro/con lists, minimum requirements, deal breakers, celebrity look-alike hopes, and on and on. As important as that question is, there's an even more important follow-up question: does that person exist?

That question reminds me of the search for extraterrestrials. "What?! How?" I hear you thinking. Well, in both cases we're looking for intelligent beings we're hoping exist that are roughly our age and have a similar understanding of the world. And in both cases these beings sometimes pop in and out of existence in the blink of an eye!

But seriously, this comparison is useful. We've spent the last half-century developing sophisticated techniques for trying to find ETs, and some of these can be adapted to help us find "the one."

One such technique dates back to the work of astronomer Frank Drake. In 1960 he used a powerful telescope to scan the radio frequencies of stars for signs of intelligent life. He found nothing. But Drake didn't give up. Instead he realized, like we just did, that searching is pointless unless you're confident what you're looking for exists. So he quantified his standards (e.g., how close to us these alien civilizations should be), used real data to estimate their values, and a year later the *Drake equation* was born.

The Drake equation estimates the number of intelligent civilizations in our galaxy using inputs like the number of inhabitable planets and the percentage of those that may be home to intelligent life. Drake and his colleagues plugged in their best estimates into the equation and got their answer—between 20 and 50,000,000 intelligent alien civilizations may exist in our galaxy [48].[1]

Let's now see if we can modify the Drake equation to help estimate how many potential partners are out there for a given person. The best way to describe this process is by illustrating it. So I'm going to do something dangerous—I'm going to pretend I'm not married and modify the Drake equation to estimate the number of women who live near me who I'd consider dating.[2] Here's the formula I came up with.

---

[1] I highly recommend you watch the YouTube clip of astronomer and renowned science communicator Carl Sagan explaining the Drake equation; the video is titled "Carl Sagan on Drake Equation."

[2] To my wife: Zoraida, I can't imagine anyone else I'd be happier with. To my toddler daughter: Emilia, daddy still loves you and mommy; this is just pretend play.

(I've inserted the multiplication symbols to avoid the equation looking like an alphabet soup.)

Estimating The Number of Potential Partners

$$N = P \times S \times A \times E \times D \times H_1 \times H_2. \qquad (5.1)$$

Here N is the number of potential partners, P is the population of the area you're looking for love in, S is the fraction of that population matching your desired sex (male, female), A is the fraction that lies within your desired age range, E is the fraction that's attained your desired level of education, D is the fraction that's amenable to dating, H_1 is the fraction of those people you'd actually consider dating, and H_2 is the fraction that would consider dating you.

Focusing my search on Boston and using the "American Fact Finder" site on census.gov, I estimate that $N \approx 350$.[*1] But then reality hits. Although 350 women is a large dating pool, it's only 0.7% of the women age 30 to 40 in Boston. I'd have to talk to about 136 Bostonian women—who I think are between 30 and 40 years old—before I could expect to meet just *one* of my special N's. (That would be a long, tiring day.) To make matters worse, my estimate is probably an *over*estimate. There are many other things I'd want that I didn't include in (5.1) (like being a Spanish speaker). These extra preferences would lower N, and one by one I'd watch my dating pool shrink. This is what happened to one young man who recently tried estimating his own N-value. He started with the nearly 4 million women that live in the London metropolitan area. As he added more terms to his own Drake-like equation, that number shrank. When he finished his detailed analysis he got the depressing result: just 26 women met all of his requirements![3]

---

[3]The *Wall Street Journal* wrote about this guy's attempt [49]; the article also mentions similar attempts by other people, including National Public Radio's David Kestenbaum (who hosts the *Planet Money* podcast I mentioned in the Chapter 4 summary).

But that's okay; just like with the aliens, all you need is $N \geq 1$. This ensures that your special someone exists. "With so much variability in the value of N, what's the point of calculating it?," you may wonder. I'll give you one good reason shortly. But for now, the real utility in this approach is that it forces you to think carefully about what characteristics you most want in a partner, and whether where you're currently looking is the best place to find that person. Then comes the fun part—dating. Once you embark on that adventure you'll (hopefully) run into another important question: when should I stop dating and "settle down"?

By now I hope you can predict what I'm going to say: "That's where math may help." I'll give you the punch line right now: under some strict assumptions (to be discussed shortly) the answer to this classic dating dilemma depends only on N. (That's why I wanted to help you estimate N, even if what pops out is a bit depressing.) We'll need to review a little probability theory to understand all of this, but I'll weave it into the description of the main result in the next section.

5.2 WHY HIRING A SECRETARY IS LIKE DATING

Let's pretend you're rich for a moment; the only problem is, you're having trouble finding someone to share your life with. So you hire a matchmaker—Mary. She signs you up for a personalized speed-dating event. Mary's found you N people she thinks are compatible with you and you're about to meet them all. (Don't worry; you look good tonight.) You know nothing about the "applicants" beforehand, but Mary assures you that you'd be able to rank them from best to worst of the bunch, with no ties, were she to give you her files on all of them (which include pictures and introductory videos).[4] So as not to overwhelm you, Mary will randomly choose one person at a time and give you two 5 minutes to chat. Sounds pretty good so far, huh?

---

[4] If you've ever seen the show *Millionaire Matchmaker*, you'll recognize my set-up.

But Mary's been doing this for a while. And she's learned that there are a few rules that should be followed. Here they are:

1. If you reject someone they're out for good; you can't go back later and contact them.
2. Once you choose your date, Mary gets rid of the remaining candidates and destroys their files; you'll never know who else you might've met.
3. You *must* pick someone. (Mary put in a lot of hard work to find you these people.)

You, of course, want to select the *best* date out of the N candidates Mary has found. Many people in this situation would do the following: interview a few candidates (but reject them), keep a mental note of the best one thus far (the "leading contender"), and then choose the first one that's better than the leading contender. (You can't simply interview them all and then pick the best one since that would violate Mary's first rule.) This strategy rejects x of the N people you interview; that's why I'll call it *the x-strategy* (in honor of the *X-Files* and the love/alien connection of the last section). The nail-biting question: what x-value yields the highest probability of selecting the best "applicant" using the x-strategy?[5]

As I foreshadowed in the last section, the answer depends on N. The $N = 1$ case is special; here there's only one candidate, and because of Mary's third rule, you're stuck dating that person. You reject no one, so $x = 0$. Since there's only one applicant they are both the best *and* worst of all the applicants. (Remember, applicants are only ranked relative to each other.) So it's only for N-values of 2 or more that we'll need to calculate the probability of selecting the best applicant.

When $N = 2$ you have a choice: pick the first person ($x = 0$), or reject them and pick the second ($x = 1$). Since we assumed you can rank candidates relative to each other, let's name these two people Average and Best. Table 5.1 shows the possible outcomes of each x-strategy. Notice that in each scenario there are two possibilities: Mary brings you Best first, or she brings you Average first. In the

---

[5] I've purposefully used the terms interview and applicant because this problem is equivalent to the secretary problem, where the goal is to hire the best secretary (see the chapter summary).

TABLE 5.1.
The $N=2$ case of the speed-dating problem. In (a) you choose the first person you meet, in (b) the second. Bold indicates who was chosen, parentheses who was never seen, and a strike-through who was rejected.

Order of Arrival	Scenario ($x=0$)	
	1	2
First	**Best**	Average
Second	(Average)	(Best)

(a)

Order of Arrival	Scenario ($x=1$)	
	1	2
First	~~Best~~	~~Average~~
Second	**Average**	**Best**

(b)

TABLE 5.2.
The $N=3$ case of the speed-dating problem. In (a) you choose the first person you meet, in (b) the second, and in (c) the third. Bold indicates who was chosen, parentheses who was never seen, and a strike-through who was rejected.

Order of Arrival	Scenario ($x=0$)					
	1	2	3	4	5	6
First	**Average**	**Average**	**Better**	**Better**	**Best**	**Best**
Second	(Better)	(Best)	(Average)	(Best)	(Average)	(Better)
Third	(Best)	(Better)	(Best)	(Average)	(Better)	(Average)

(a)

Order of Arrival	Scenario ($x=1$)					
	1	2	3	4	5	6
First	~~Average~~	~~Average~~	~~Better~~	~~Better~~	~~Best~~	~~Best~~
Second	**Better**	**Best**	Average	**Best**	Average	Better
Third	(Best)	(Better)	**Best**	(Average)	**Better**	Average

(b)

Order of Arrival	Scenario ($x=2$)					
	1	2	3	4	5	6
First	~~Average~~	~~Average~~	~~Better~~	~~Better~~	~~Best~~	~~Best~~
Second	~~Better~~	~~Best~~	~~Average~~	~~Best~~	~~Average~~	~~Better~~
Third	**Best**	**Better**	**Best**	**Average**	**Better**	**Average**

(c)

$x=0$ strategy you choose Best in one out of two scenarios (bold shows who was chosen and parentheses who was never seen); that's a 50% probability of selecting the better applicant. Let's record this as $P(2, 0) = 50$, where the first number in the parentheses is N and the second x. The $x=1$ strategy also has the same odds: $P(2, 1) = 50$ (the strike-throughs indicate who was rejected).

Things get more interesting when $N=3$. Let's once again name these three people according to your relative ranking of them: Average,

TABLE 5.3.
The second column shows the optimal x-value (number of people to reject) in the speed-dating event. The third column expresses that number as a percentage of the total number of candidates (N). The last column gives the probability of selecting the best candidate using the x-strategy. Values have been rounded.

N	Optimal x	x/N (%)	$P(N, x)$ (%)
3	1	33	50
4	1	25	49
5	2	40	43
6	2	33	43
7	2	29	41
8	3	38	41
9	3	33	41
10	3	30	40
11	4	36	40
12	4	33	40
13	5	38	39
14	5	36	39
15	5	33	39

Better, and Best. Table 5.2 shows the new possible outcomes. Notice that in both the 0-strategy and the 2-strategy (panels (a) and (c), respectively) you choose Best in only two out of the six scenarios, so that $P(3, 0) \approx 33$ and $P(3, 2) \approx 33$. But check out the 1-strategy (panel (b)). In Scenario 3 you meet Better first. They are by default the leading contender, and since $x = 1$ you reject the candidate. Then Mary brings you Average. But they aren't as good as the leading contender (Better), so you decide to hold out for the next one. And wham! Best walks in! So the 1-strategy selects Best in *three* out of the six scenarios, making $P(3, 1) = 50$. That's higher than the 0-strategy and the 2-strategy, so for $N = 3$ the 1-strategy is the way to go; you'll select the best applicant 50% of the time by following it.

For $N = 4$ there are 24 possibilities. I'll spare you the corresponding tables and jump straight to the results. Table 5.3 shows the optimal N-value (first column), x-value (second column), x as a percentage of N (third column)—for example, in the $N = 3$ case the optimal strategy is to reject the first 33% of candidates—and the probability of selecting the best candidate using that x-strategy (fourth column). Here are two noteworthy observations.

- The probability of selecting Best, $P(N, x)$, decreases as N increases.
- *But* notice that as N gets larger the third and fourth column numbers seem to settle down in the 30s. Coincidence? No way; it's math! One can *prove* mathematically that *as N tends to infinity the optimal strategy is to reject the first 37% of candidates and then choose the first one that's better than the leading contender. Moreover, this strategy selects the best candidate at least 37% of the time.*[6]

The results above are based on the assumptions I presented (which include Mary's three rules). Real life will rarely match those assumptions; for starters, you probably don't have a "Mary" to bring you potential dates, and you can certainly go back and contact people you originally rejected. Without those assumptions the 37% result doesn't necessarily follow.[7]

Nonetheless, there are a couple of takeaways from this analysis. The main one: it's just as important to think about the *selection* process we use to find a partner. (The x-strategies underlie the 37% rule.) This complements the takeaway of the previous section, which got us thinking about the *search* process. Second, the math tells us that the optimal x-value depends on N. At the end of the previous section you might have thought that having a large N is better than having a small N. But look back at Table 5.3—as N increases, the probablity of selecting the best candidate *decreases* (at least if we're using x-strategies to make the selection). So maybe that picky British guy from the previous section was on to something—perhaps selecting from a small pool of highly compatible candidates is better than selecting from a large pool of overall less compatible candidates.

Even though the assumptions in the set-up leading to the 37% rule are not realistic, isn't the result amazing? We're talking about dating, about love, and yet math uncovers hidden insights. "But 37% is a fairly low probability," you may say. I agree. But this isn't a deficiency of the

---

[6]The actual percentage (in decimal form) is $1/e \approx 0.3678$ (see [50] for a proof); here $e \approx 2.71$ is *Euler's number*. A more advanced approach to this problem allows you to completely disregard N. But this approach requires that you make assumptions about how often you run into people you'd consider dating (see [51], Example 2).

[7]Incidentally, the 37% rule also shows up in a variety of other situations; see the chapter summary for a short discussion.

mathematics. It's just the best we can do given what was assumed. If we had more information we might be able to do better. Indeed, that's what I'll show you in the next section. I'll take you to a more typical speed-dating event, this time involving N men and N women, and show you how more information (plus math) can create couples whose members won't cheat on each other.

5.3 THE STABLE MATCHING PROBLEM

Let's bring back Mary the matchmaker. She's invited you to another dating event, this time involving N men and N women, all hetero-sexual.[8] Before the event happens she mails you a big box—inside are her files containing *complete* information on *every* member whose sex is opposite to yours. There's also a note: "Please study these files and rank these applicants from best to worst. Bring this list with you to the event."

The night of the event your chauffeur drops you off at the venue: a large ballroom. When you walk inside you're given a nametag by one of Mary's assistants. The ballroom is nowhere near as glamorous as you were expecting. There's a long rope on the floor dividing the room in half, and that's it. On one side is a sign that says "Gentlemen here," and on the other one that says "Ladies here." Everyone lines up on the appropriate side as Mary emerges from the back of the room.

"Good evening, everyone; thank you for coming. I'm sure you're wondering what this rope is about, and I'll explain that in a minute. For now, please put your names on your name tag and take out your preference lists."

Mary, it turns out, mailed *everyone* her complete files on attendees of the opposite sex. Those boxes contained her same note to you. So everyone now has a preference list that ranks members of the opposite sex in order from best to worst.

"Okay, let's get this event started. Here's how it's going to work. I'm going to lead an unspecified number of matchmaking rounds in

---

[8]I discuss the heteronormativity assumption later.

which each unmatched man will ask out a woman.[9] When everyone is matched up and no two people would rather be with each other than with their current partners, we'll end the event."

In math terms, Mary's trying to accomplish *stable matchings*. What she doesn't mention is how she'll go about matching people (that's her secret). But she's a smart woman, so she's going to use math to help guarantee that all matches are stable. Let me describe her method, and then I'll give you some takeaways for our own dating lives.

"Is everyone ready?," yells Mary. "Okay, let's begin. Men, please cross the rope and ask out your top choice." We can all guess what happens next: a swarm of men walk over to one particular woman while the others scatter among the rest. At the end of this first round, each woman has received either no proposals, one proposal, or multiple proposals.

"To those of you that haven't gotten a proposal yet, just hang in there, they're coming" says Mary. "To the others, please do the following: if you've received only one proposal, accept it and consider yourself 'engaged'; if you've received multiple proposals, accept the guy highest on your list and consider yourself engaged."

At the end of this first round there are still men and women who don't have a partner. Mary asks the unengaged men to fall back behind the rope. Then, she starts round two. "For this second round, all unengaged men, please go ask out your *second* highest choice."

Here's where things get awkward—some men are asking out women that are already engaged. But Mary's got a solution for that; she quickly intervenes to prevent chaos.

"Okay, ladies: if you're engaged and are receiving proposals, feel free to dump your current man and *trade up* if the second-round man is higher up on your preference list."

Ouch! (But then again, the men chose first in round one.) Mary senses the rejection some people feel in the room and she tries to keep the morale up.

"Men: if you were dumped, don't take it personally. For you all and whatever other men remain unengaged, in this next round please propose to the woman you are most interested in that you haven't

---

[9] I'll discuss the assumption that men choose first later.

proposed to. Ladies: once again please accept if you receive just one proposal, pick your top choice if you get multiple ones, or feel free to trade up if you're currently engaged. If you haven't been asked out yet, I promise you will be."

Mary continues in this fashion round after round. It takes a while, but eventually something magical happens: everyone's matched up and all matchings are stable! Mary's secret is that she was creating couples according to the *Gale-Shapley (GS) algorithm*, named after the mathematicians who solved the *stable matching problem* in 1962. They proved that the GS algorithm accomplishes Mary's two goals: get everyone matched up and have every couple be stable. I guide you through the proof of the first claim in the appendix.*[2] As for the stability of the matchings—the real secret sauce in Mary's work— let me discuss the (short) proof that the GS algorithm accomplishes that.

Consider two people at the event: Jessica and Jorge. Jessica is an engaged woman, Jorge an engaged man, though they're not engaged to each other. But not all is well—Jorge prefers Jessica to his current partner. That's okay, because the GS algorithm Mary used guarantees that Jessica *does not* prefer Jorge to her current partner. Here's why. Since Jorge prefers Jessica over his current partner he must have proposed to Jessica *before* proposing to his current partner. If Jessica rejected his proposal, she was already engaged to someone else she preferred more than Jorge. Takeaway: Jessica prefers someone else over Jorge. Now let's go back and suppose that Jessica accepted Jorge's proposal. Well, she's not currently engaged to him, so she must have dumped him somewhere along the way when she traded up. Takeaway: Jessica prefers someone else over Jorge. In either scenario, even though Jorge wants to be with Jessica, she's already with someone better and is unwilling to leave him for Jorge. Pretty cool, huh?

This short proof also illustrates one amazing feature of the GS algorithm: it prevents cheating (or, in the case of marriage, affairs). That's because even though there may be a "Jorge" in the pool of men that "covets his neighbor's wife" (some other woman, "Jessica"), the algorithm ensures that "Jessica" prefers to be with her current partner rather than "Jorge."

There are downsides to the GS algorithm though. Here are the three worst ones:

- As we just saw, some couples may include one person that would rather be with someone else, even though that person is happier in their current relationship. So, although the GS algorithm prevents cheating, it doesn't prevent people from *wanting to cheat*. (There's only so much we can ask of math.)
- The assumption of heteronormativity is *necessary* to get stable matchings. The similar problem of matching roommates, which doesn't assume heteronormativity, doesn't always have stable matchings.
- There may be multiple stable matchings. This wouldn't be so bad except for this fact: the algorithm *guarantees* that out of all the possible stable matchings, each man ends up with the *best* partner he can have while each woman ends up with the *worst* partner she can have (based on the preference lists) [52].*3

That last one is particularly troubling. But one of the great things about math is its universality. In this context, it means that if Mary swapped the roles of men and women at the event and had the *women* choosing in every round, then *they* would end up with their best possible match, and the *men* would end up with their worst possible match. The takeaway in the words of Mary the matchmaker: *ladies, go get 'em!*[10]

From searching for aliens to preventing cheating, this chapter has been the most diffuse of the chapters. But what did you expect? We're talking about love here; this stuff is complicated! Despite that, mathematics has provided us with some profound results. Some of the assumptions needed to derive those results are unrealistic, but we learned a lot by working through the derivations. The 37% rule and the cheating-proof GS algorithm stand out in particular. The striking realism of these results still amazes me.

I've saved the best for last, though. In the next chapter I'll show you how researchers have used math (and science) to predict the outcome of our relationships (e.g., happy ever after). I'll also give you

---

[10]There are other neat (and useful) applications of the GS algorithm; see the chapter summary for a short discussion.

Dr. Math's recipe for making joint decisions that are perceived as fair and transparent by both parties in a relationship. Finally, I'll discuss some promising research that may help prevent a break-up.

Chapter 5 Summary

MATHEMATICAL TAKEAWAYS

- The mathematics of probability can be used to describe the likelihood that something will happen. In the simplest case, the probability of event X happening is the ratio of the number of ways that X can occur to the total number of possible outcomes. For example, if X is "the probability of rolling an even number on a fair, 6-sided die," then since there are 3 even numbers on the die (2,4,6) and 6 possible outcomes, $P(X) = 3/6 = 0.5$ (i.e., a 50% chance of rolling an even number).

- Probabilities can range from 0% (the event never happens) to 100% (the event always happens).

- The 37% rule arose historically out of the "secretary problem." In this problem an employer must interview candidates for a secretary position one by one and decide on the spot whether to hire or reject each person. Any hiring decision—whether it be secretaries, teachers, cops—that satisfies the assumptions of the secretary problem is amenable to an x-strategy approach; see Table 5.3 (for small N) or the 37% rule (for larger N) for the optimal x-value.

- The Gale-Shapley (GS) algorithm, like the secretary problem, has many other applications. In fact, in the article in which Gale and Shapley presented their algorithm they also studied college admissions [53]. Moreover, the National Residents Matching Program has used variants of the GS algorithm since 1952 to match medical school students with residency programs at hospitals; see [54] for yet other applications (and generalizations) of the stable matching problem.

NONMATHEMATICAL TAKEAWAYS

- Any Drake-like approach to estimating your number of potential partners is only as good as the assumptions it makes. Make too few assumptions and you'll get a large pool of "candidates." Make too many assumptions and you might end up with just a handful. Regardless, the real utility in this approach is that it forces you to think carefully about what characteristics you want most in a partner. Once you know that, you can use other tools—like online dating sites, event calendars, or even demographic data—to help increase the chances that you'll run into one of those precious few N's.

- The x-strategies of Section 5.2 are similar to the dating approach many of us already use. When all of the assumptions discussed in that section are satisfied, the 37% rule—which suggests we reject the first 37% of the people we meet and settle down with the next person who's better than the leading contender—guarantees at least a 37% chance that you settle down with the best possible partner. That's not a very high probability, and the assumptions needed to guarantee that number are unrealistic, but that's the best the math can do given the few broad assumptions made.

- The GS algorithm can be used to construct heterosexual stable couples—couples whose members won't cheat on each other—assuming each person has a preference list that ranks all candidates from best to worst, that there are as many men as there are women, and that the men choose first in every "matchmaking" round.

- Unfortunately, the stable couples created by the GS algorithm are only guaranteed in the heterosexual scenario. Additionally, whatever sex does the proposing ends up with their best possible match out of all the possible stable matchings, while the other sex ends up with their worst possible match out of all the possible stable matchings.

BONUS: A FEW PRACTICAL TIPS

- *Have fun with it*. Dating is an adventure. The results in this section aren't meant to be applied directly, but to get us thinking about how we go about that adventure.[11] Then again, if you know someone who runs speed-dating events, teach them the GS algorithm and beg them to use it, just once. Then email me; I'd *love* to hear about the results.

---

[11] For instance, I'm no relationship expert, but I'd guess your dates wouldn't appreciate being rejected *immediately* because you've yet to reach your 37% threshold.

CHAPTER 6

■■■■■■■■■■■■■■■■■

Living Happily Ever After with "The 1"

OKAY, YOU'VE FOUND SOMEONE you think might be a keeper. This person shares your most important values, you have common interests, and your best friend and family approve. You've only been texting and emailing, though; you've yet to meet face to face. So one of you arranges a low-stakes meeting at a coffee shop ...

You arrive first. Your date—let's call that person X—walks in the door a few minutes later. What happens the instant you see X is a mystery to all of us; it's a mix of psychology, chemistry (both the science and the "chemistry" between two people), biology, neuroscience ... you name it! But from my mathematical perspective the two of you are now a *dynamical system*. Such systems are characterized by two or more things—the "system," in this case two people—interacting over time (the "dynamics"). (For example, you smile, X smiles and nods, then you fix your hair, and on it goes.) Mathematicians have spent hundreds of years studying dynamical systems. Part of what we've discovered[1] is that they contain hidden properties that help demystify the dynamics regardless of however complicated they may be. Sounds like just what we need. That's why I'm going to spend this chapter discussing "relationship dynamics" from this viewpoint.

We'll start off by describing a realistic dynamical systems model of a romantic relationship. I'll tell you about the suggestions this model makes for how to strengthen a relationship. (Yep, even a mathematical model gives relationship advice these days.) It's *good* advice, and

---

[1] My own research is in a subfield of dynamical systems.

somewhat surprising, too: one of the results is the partitioning of couples into two groups based on their ability to deal with shocks to the relationship. *Robust* couples can deal with shocks and still recover (e.g., finding out one person has developed feelings for someone else), while *fragile* couples can only survive shocks if they're not too intense. (I told you this model was realistic!) But fear not, I'll also discuss recent research findings that may help couples keep their relationship intact. We'll also learn some new mathematics derived from a field called *game theory* and use it to derive equations that can help couples make joint decisions perceived as fair and transparent by both people. So let's get started.

6.1 YOUR RELATIONSHIP AS A DYNAMICAL SYSTEM

Let's go back to the instant you meet person X for the first time. To mathematize the interaction, let's introduce the following variables and parameters.

1. Let's label the intensity of your feelings for X by y (y for "your"), X's feelings for you by x, and let's also say that *positive* values indicate liking the other person, while *negative* values indicate *dis*liking the other person. (For example, $y = 10$ indicates that you like X.)
2. The values of x and y will change throughout your date. So, let's call y' (read "y prime") the *instantaneous* change in your feelings about X, and x' the instantaneous change in X's feeling about you. Realistically, we all have noninstantaneous reaction times, but this assumption keeps the math simpler.
3. The instant you see each other you both form the first impression. Let's denote by A_x the appeal of X to you, and by A_y the appeal of you to X. You two already know a lot about each other (you've been texting and emailing, remember?), so A_x and A_y quantify not only physical attractiveness but also other desirable characteristics like socioeconomic status or cultural heritage.

There are lots of other potentially relevant factors, but I hope you agree that these variables/parameters describe the core of the initial interaction between you and X. All we have to do now is figure out how x' and y' are related to x, y, A_x, and A_y. Let's do that by following the scene second by second.

We start with the first impression. This certainly has an effect on both of your feelings for each other. So the equations for x' and y' should contain A_y and A_x, respectively. But we don't know the exact effect. So, let's call $f(A_x)$ the function that describes the effect of X's appeal on your feelings for X.[2] Similarly, let $g(A_y)$ be the function describing the effect of your appeal on X's feelings for you. Think of f and g as "interest functions" that describe how interested one person is in the other based on the appeal of the other person.[3] We can then write down our first guess at the equations for x' and y':

$$x' = g(A_y), \tag{6.1a}$$

$$y' = f(A_x). \tag{6.1b}$$

These equations say that the instantaneous change in one person's feelings for the other (i.e., x' and y') is equal to their interest in that person (the function f or g).

X now sits down and you two start talking. Like Newton's third law, every action leads to a reaction (though not necessarily equal and opposite). As you two converse, you pick up on each other's subtle cues containing indications of your feelings for one another. Let's encode this with the functions R_x and R_y—your reaction to X's feelings for you is $R_y(x)$ ("your reaction to x's feelings") and X's reaction to your feelings for X is $R_x(y)$. Our updated model becomes

$$x' = g(A_y) + R_x(y), \tag{6.2a}$$

$$y' = f(A_x) + R_y(x). \tag{6.2b}$$

---

[2] The notation $f(A_x)$ is read "f of A_x" and is shorthand for "function of A_x."

[3] You could equate "interest" and "appeal" if you wanted (just choose $f(A_x) = A_x$ and $g(A_y) = A_y$), but I think of them as being different. Here's a thought experiment to support my case. Let X be an actor you think is attractive. Now, imagine two versions of you: version A just got out of a long-term relationship with a look-a-like because that person cheated on you, while version B has been single for a year. I'd bet that A's *interest* in X would be lower than B's, despite X having the same *appeal* in both scenarios (e.g., socioeconomic status, good looks).

This model now incorporates the initial interest level and the ongoing reaction to each other's feelings.

Finally, if we fast-forward for a moment to the end of the date, right after X leaves you'll still be thinking about them. But as the day progresses you'll get back to thinking about your normal life. Effectively, this means the intensity of your feelings for X will decrease a bit. Let's assume this decrease is exponential at the rate d_y, while X's feelings for you decrease exponentially at a rate d_x (both positive numbers).[4] We incorporate these two extra assumptions by adding the terms $-d_x x$ and $-d_y y$ to (6.2a) and (6.2b), respectively. (The reason for this has to do with calculus; check out the appendix for a short explanation and introduction to calculus.)[*1] That brings us to the final version of the model:

$$x' = g(A_y) + R_x(y) - d_y y, \qquad (6.3a)$$

$$y' = f(A_x) + R_y(x) - d_x x. \qquad (6.3b)$$

I doubt we'll ever find the exact functions f, g, R_x, and R_y. But what we can do is to make simple, realistic assumptions about them. (For example, we can assume that your interest in X increases as X's appeal increases, which implies that the values of $f(A_x)$ increase as A_x increases.) This is what two mathematicians did in 1998 when they analyzed our dynamical system [55]. The results give even the best couples therapists a run for their money.

The first thing they found is that this model has three *equilibrium states*, states where each person's feelings for the other don't change. Moreover, they found that the couple, over time, drifts toward one of these equilibrium states.

But which state a couple eventually ends up in depends on another one of their findings: couples come in two varieties—robust and fragile. Robust couples' feelings for each other (x and y) get increasingly positive with time. They drift toward their *only* equilibrium state: a happy state where each person likes (maybe even loves) the other.

---

[4] In Chapter 3 we saw how quickly exponential functions change in value, so you might question this assumption. But note that we get to control the rate of change via the parameters d_y and d_x; if they're tiny numbers, the decrease is gradual.

Moreover, this happens regardless of their initial feelings for each other. Fragile couples aren't so stable. They have three equilibrium states, and which one they end up in depends on their initial feelings for each other. If they weren't too negative, their relationship will evolve into a happy state; if they were, they'll eventually end up in one of two unhappiness states. (See the appendix for a discussion of how these results are derived from (6.3).)^{*2}

A particularly interesting outcome of the model is this: (6.3) contains three hidden pieces of advice that can help make "fragile" relationships more like the "robust" ones:

1. **High enough interest on both ends ensures robustness.** If you and X are sufficiently interested in each other, the model predicts an eventual equilibrium state of happiness. This echoes how important compatibility is in selecting a partner.[5]

2. **Blunting the effect of shocks to the relationship can avoid breakups.** A fragile couple may be thrown into an unhappy state and be doomed to break up if they suffer a severe enough shock. The "severe enough" qualification comes from another feature of the model: it contains a threshold that separates fragile couples' two unhappy equilibrium states from the single happy one. So if the couple works on increasing that threshold, they'll be better able to absorb shocks and not cross the threshold. The model then predicts the relationship will eventually find its way back to the happy equilibrium state.

3. **Improving the appeal of one individual leads to more positive feelings for *both* people at equilibrium.** Translation: each person's work on improving themselves—getting healthier, increasing their financial stability, and so on—increases the love potential in the relationship.

Though this is a far-from-complete model of "love dynamics," I hope you're as impressed as I am with this list of relationship advice, especially given that it came from a mathematical model! Notice too the harmony between this model's advice and what we learned in the

---

[5]It's also another argument for broadening the definition of appeal beyond just looks. (Hollywood couples are notorious for having good looks *and* short relationships.)

earlier chapters. To ensure high-enough interest on both ends, for example, both people in the model should have a very good idea of who they're looking for. This sounds exactly like the argument I made in the previous chapter for creating your own Drake-like equation. Similarly, you just read four chapters packed with strategies for improving one's health and finances, and would you look at that, the third piece of advice above suggests that implementing that knowledge could strengthen relationships! (This mathematical model even has something to say about the stable matching problem.*[3]) Finally, let me point out that the assumptions of this model are general enough that the results apply to nonromantic relationships, too (for example, the working relationship between an employee and their boss). In many ways, then, our calculus-based model is, more generally, a model for happiness (cue the title of this book).

Since the earlier chapters address this model's first and third pieces of advice, I want to focus next on the second piece. In the next section I'll show you how math can improve the communication between you and your partner and in the last section tell you how it can help you deal with conflict in the relationship (including avoiding breaking up). But first we'll have to once more draw on new math to help us out.

6.2 NEED HELP MAKING A JOINT DECISION? THERE'S AN EQUATION FOR THAT

Here's a question for you. Imagine you get a $500 check in the mail from some class-action lawsuit. Your partner is with you at the time, and the two of you need to decide what to do with the money. You, having read Chapter 3, are inclined to *save* it ("think how much sooner we could retire," you argue). But your partner wants to *spend* it ("think of how much fun we could have tonight," your partner responds). You two decide to divide up the $500. How much should each person get?

Well, there's always the easy way out: each person gets half. But that may leave you (or your partner) dissatisfied. Wouldn't it be nice if there was a fair and transparent way to solve this problem? You're in luck, because there is.

In 1950 a 22-year-old man named John Nash found one. He devised a mathematical method to make these kinds of joint decisions and proved mathematically that his approach results in a fair, and to a degree, "cheat-proof" agreement between the two people.[6]

You may remember the name John Nash from the Oscar-winning 2001 film *A Beautiful Mind*, in which Russell Crowe portrayed Nash, a brilliant mathematician who practically created a new branch of mathematics called *game theory*. Nash did his early work on "noncooperative games." By "games" he meant any sort of interaction between rational people ("players") that involved making decisions that affected other players' "payoffs." For example, poker qualifies as a (Nash) game, and so does two countries' decision to go to war. Lucky for us, Nash was also interested in *cooperative* game theory. His paper "The Bargaining Problem" [57] is where our decision-making equation can be found.

In the paper Nash studies situations in which "two individuals … have the opportunity to collaborate for mutual benefit in more than one way." He then adds the assumptions that "no action taken by one of the individuals without the consent of the other can affect the well-being of the other one," and that "the two individuals are highly rational, that each can accurately compare his desires for various things, that they are equal in bargaining skill, and that each has full knowledge of the tastes and preferences of the other." Maybe the assumption of *full* knowledge of "tastes and preferences" is a bit much, but otherwise Nash's assumptions are fairly realistic.

Let me describe how to adapt Nash's approach to our context to produce one equation that you and your partner can use to make optimal joint decisions about nearly anything. Here are the steps.

> *Step 1: mathematize happiness.* Nash's approach relies on the concept of *utility functions*. These functions quantify you and your partner's preferences over the possible outcomes. For our purposes, think of these functions as quantifying the happiness you get from the outcome of an agreement. Let's label your utility function $Y(x)$ and agree that it'll measure the happiness you get from saving x of those $500. Similarly, let's label your partner's utility function $P(z)$;

[6]See [56] for a discussion of the fairness and other similar characteristics of Nash's approach.

it measures their happiness from spending z of those $500. Let's also assume four more things:

- Happiness is measured on a scale from 0 to 10, with 10 representing "happy" and 0 "unhappy."
- You get more happiness than your partner as your share of the $500 increases.
- Both of you derive zero happiness from receiving no money.
- If you get all $500 your happiness is at a 10; if your partner gets it all, their happiness only reaches an 8.

There are many functions that satisfy these assumptions. But I'll keep things simple and give you the *linear* ones (see the appendix for a verification that the following functions satisfy the assumptions):*[4]

$$Y(x) = \frac{10}{500}x, \qquad P(z) = \frac{8}{500}z. \qquad (6.4)$$

Step 2: determine the constraints. We've already encountered three constraints:

$$x + z = 500, \qquad x \geq 0, \qquad z \geq 0. \qquad (6.5)$$

The first says that whatever you get and whatever your partner gets must add up to $500; the last two indicate that you each get at least $0.

Step 3: mathematize disagreement. Next we need to quantify how you'd both feel in the event no agreement can be reached. Let's assume that in this case you'd be left at a happiness level of 3 (bummed) and your partner at a 4 (not so bummed).

Step 4: maximize the Nash product. Nash's optimal agreement is given by the solution to this problem:

$$\text{maximize } (Y - 3)(P - 4), \qquad (6.6)$$

where we restrict the search for the maximum to all pairs (Y, P) such that $Y \geq 3$ and $P \geq 4$ (i.e., we only consider possibilities that are no worse than the "no agreement" scenario); the product $(Y - 3)(P - 4)$ in (6.6) is called the *Nash product*.

I solve this problem in the appendix by finding the maximum of a quadratic function;*[5] the answer is to give your partner $300 and allow yourself to save the remaining $200. This solution leaves you at a utility value of $Y = 4$ and your partner at a utility value of $P = 4.8$. So it leaves your partner with a larger share of both the money and the happiness. "But saving matters more to me than spending does to my partner," you say. True, but Nash's solution is fair, remember? It doesn't seek to maximize any one person's utility.[7] I'll come back to this in a minute. First let me give you the general formula so you can plug in your own numbers and get a math-backed optimal decision-making equation.

Calculating Optimal Joint Decisions

Suppose you and your partner need to decide on how to divide up a total T of something (e.g., money). Let M denote your happiness level were you to receive *all* of that something and N your partner's, and suppose happiness is measured on a scale from 0 to 10 (with 10 representing "happy" and 0 "unhappy"). Finally, let Y_d and P_d be the happiness levels you and your partner, respectively, would experience in the event no agreement can be reached. Then the optimal agreement is to divide up T as follows:

$$\text{Your share} = \frac{T}{2}\left(1 + \frac{Y_d}{M} - \frac{P_d}{N}\right), \qquad (6.7a)$$

$$\text{Your partner's share} = \frac{T}{2}\left(1 + \frac{P_d}{N} - \frac{Y_d}{M}\right). \qquad (6.7b)$$

---

[7] That's why there's a product in (6.6); that takes into account *both* utility functions in the maximization process.

I derive these equations in the appendix,*6 but right now I want to mention a few things. First, let me point out that T could be anything divisible; it doesn't have to be money, it could also be time (say you want to stay home and your partner wants to go out) or food (e.g., slicing a cake or pizza). I also want to discuss how to "game the system," and how, in a surprising twist, whoever succeeds will have strengthened the relationship! I'll show you what I mean by considering a few special cases of these equations.

- For starters, notice that if the disagreement payoff for both is zero (i.e., $Y_d = P_d = 0$) then (6.7a)–(6.7b) give the same number: half of T. So if you're both unhappy with a disagreement, this process suggests you two just split T evenly and call it a day. I like this because it's math looking out for both of you. It's like it senses the potential danger of a heated argument as you both spiral toward that "unhappy-unhappy" utility combination, and it protects you both from reaching that point by saying "let's just split T and avoid any problems."

- Consider now the case when $Y_d = P_d$ but they're not zero. Here disagreement doesn't produce unhappiness (though there may be disappointment). In this case it turns out that if $N > M$ (i.e., your partner's maximum happiness is *greater* than yours) then your partner's share of T will be *smaller* than yours.*7 I like this conclusion too; it illustrates the fairness inherent in (6.6).

- Finally, consider the case when $M = N$, which happens when the maximum happiness is the same for both people. Then if $Y_d > P_d$ (i.e., your disagreement payoff is greater than your partner's) *you* get a larger share of T than your partner.*8 For your partner to get a larger share than you, we'd need $P_d > Y_d$. In other words, your partner would have to be *happier* than you in the event no agreement can be reached.

Aren't these three properties amazing? Especially that third one, which as I hinted at earlier, suggests one way to "cheat" Nash's mathematics: get yourself to feel happier about the possibility that no agreement will be reached. That's a positive thing! Wouldn't you rather end a disagreement by saying, "oh well, we couldn't agree; let's go out for ice cream," than one (or both) of you walking away angry? Plus, if you manage to do this (6.7a) rewards you with a larger share of T! How's that for incentive to look on the bright side of things? Finally, I want to remind you that (6.7a)–(6.7b) have a computer icon next to them, so feel free to use the customizable online versions I've put on the book's website.

Now, there are bound to be times when even math can't protect you and your partner from having negative feelings toward each other. But it turns out that researchers may have found a relatively simple way to minimize the effects of this. The idea goes back to our work in the previous section: view the interaction—likely in this case an argument between the two of you—as a dynamical system and use math to discover actions that can drive the system toward a happier equilibrium. I'll tell you that story in the next section.

6.3 HOW PSYCHOLOGISTS USE MATH TO PREDICT DIVORCE

In 1999 psychologists John Gottman and Catherine Swanson and mathematician James Murray published an influential study that effectively mathematized divorce. They reported on the results of filming 130 newlywed (heterosexual) couples having 15-minute discussions of heated topics (e.g., politics) and then using math to analyze the results and find patterns. Surprisingly, they were able to predict with high accuracy whether a couple was divorced or happily married 6 years later [58]. Conveniently for us, the mathematical model they used was very similar to the dynamical system (6.3). Here's their model adapted to our notation:

$$W' = a + bW + R_W(H), \tag{6.8a}$$

$$H' = c + dH + R_H(W). \tag{6.8b}$$

Here H and W effectively represent the husband's or wife's happiness throughout the conversation (negative values indicated negative feelings, and positive values positive feelings),[8] a, b, c, and d are numbers, and $R_W(H)$ is the function that encodes the wife's reactions to what her husband is saying, while $R_H(W)$ encodes the husband's reactions to what his wife is saying.

The similarity between the systems (6.3) and (6.8) means that (6.8) can be analyzed in the same way (6.3) was. But Gottman's team did something different: they chose a specific form for the $R_W(H)$ and $R_H(W)$ functions—a piecewise linear function made up of three horizontal (i.e., zero-slope) lines. They called the middle line's boundary points the *negativity threshold*, defined by them as "the point at which negativity has an impact on the partner's immediately following behavior." They then inserted the data from the 15-minute discussions into the updated equations (6.8) and correlated the negativity thresholds they got with the outcomes of the marriages. The rule of thumb they found predicted which couples eventually divorced with *94% accuracy* [59]: *couples with low negativity thresholds, where spouses "are noticing and responding to negative behavior when it is less escalated," had the highest chances of having a happy, healthy marriage.*[9] This makes sense; in relationships like these, negative feelings are prevented from snowballing into larger problems later. Yet what stands out to me is that once again it's a *dynamical system* that's producing sound relationship advice. (I hope this is making you want to learn calculus!)

To be fair, these results have their limitations. For starters, other researchers have criticized the team for drawing conclusions based on too small a sample size (the 130 couples in the study) [60]; I'm skeptical too of that 94% number.

On the other hand, work using mathematical modeling alone (no heated discussions) seems to support the overall idea of a low negativity threshold. For example, in [61] mathematicians used a dynamical systems approach similar to (6.8) (with similar assumptions) to study how the *delay* in one person's reaction to negativity affects the relationship. The researchers found a "Goldilocks" zone—*too quick or too delayed a*

---

[8]The researchers assigned values to these variables by quantifying everything from facial expressions to vocal tone.

[9]Gottman's book (see [52]) also discusses the success of his team's approach when applied to same-sex couples, in addition to studying the effect of a baby on a marriage.

response could eventually destabilize the relationship. Put this together with the negativity threshold found by Gottman's team and we arrive at an intriguing possibility: *perhaps relationships are made stronger by each person having a low tolerance for negative behavior and each letting the other know (fairly quickly) when such behavior arises.*[10] Of course, these conclusions are based on mathematical models and/or social science, whose assumptions may not hold for particular couples, so their conclusions shouldn't be taken as gospel.

From fragile and robust couples to Nash—who sadly passed away in 2015 (though he lived long enough to enjoy his 1994 Nobel Prize in Economic Sciences and his 2015 Abel Prize)—and game theory, from bargaining as a mathematical problem to a way of making fair and transparent joint decisions in a relationship, from pondering the benefits of a low negativity threshold and fast-ish response to negativity to discovering that some dynamical systems preach the same relationship advice as the Bible. This chapter's been quite the adventure huh?

I hope this last chapter has convinced you (if you weren't already) that mathematics can be useful, insightful, and understandable all at once. In the Epilogue I'll share my bird's-eye view of what we've covered in this book. I'll also ask you to help me spread the love of mathematics to your friends and family, including to those little ones that are "eager young minds" (in the words of Russell Crowe in *A Beautiful Mind*). Hopefully what you've learned in this book will help you make the case that the question is not "why *should* you study mathematics?"; the question is "why *shouldn't* you?".[11] And to further encourage *you* to continue learning math, here's what a couple of well-known thinkers had to say about that:

> "If I were again beginning my studies, I would follow the advice of Plato and start with mathematics."—Galileo Galilei

---

[10] This relationship advice is also backed by a higher power—the Bible: "do not let the sun go down on your anger" (Ephesians 4:26).

[11] This is a reference to a famous scene in the 1998 movie *Good Will Hunting*, a movie about a math genius.

Chapter 6 Summary

MATHEMATICAL TAKEAWAYS

- Dynamical systems model the evolution (in time) of interacting "agents" (e.g., people, countries). Typically, these systems are modeled using equations that involve instantaneous rates of change—this concept is covered in calculus courses, where it's called the *derivative*—as well as functions that describe the current state of various characteristics of each agent. As time passes the agents in a dynamical systems model find themselves in different states. In some systems there are equilibrium states—states where no further change is experienced by the agents.
- Game theory is a broadly applicable branch of mathematics whose methods and results give insights into virtually any decision-making process.
- The Nash bargaining problem with linear utility functions (what I presented in Section 6.2) is an optimization problem that can be solved by finding the maximum of a quadratic function (the Nash product).

NONMATHEMATICAL TAKEAWAYS

- In the dynamical systems view of a relationship discussed in Section 6.1, couples come in two varieties: robust or fragile. Robust couples eventually have a happy relationship, regardless of their initial feelings for each other. Fragile couples only end up in a happy relationship if their initial feelings for each other weren't too negative.
- One way to ensure robustness is for both people to be sufficiently interested in each other. In the model this is based on each person's appeal, defined to include not just physical attractiveness but also other possibly relevant characteristics (like level of education).
- Couples—fragile or robust—can also increase their relationship's love potential by increasing one (or both) person's appeal.
- Shocks to the relationship may cause fragile couples to tend toward an unhappy state. But this only happens if the shock is severe enough to cross the happy/unhappy threshold. Couples in the model can

protect their relationship from the effects of shocks by increasing this threshold (i.e., finding ways to cope with more severe shocks).

- One of the more practical takeaways from the chapter is the Nash bargaining problem solution (see Section 6.2). It provides a fair and transparent way to make joint decisions.
- Another practical takeaway is that preventing conflict in the relationship from snowballing into a larger problem may protect against break-up (or divorce). Gottman's research, as well as evidence from other sources, suggests that having a low negativity threshold—where each person doesn't wait too long before voicing their dissatisfaction or unhappiness—could help insulate a couple from break-up/divorce. Results from a similar mathematical model raise the interesting possibility that perhaps not waiting too long before bringing up your dissatisfaction may also strengthen a relationship.

BONUS: A FEW PRACTICAL TIPS

- *Learn more about these topics on YouTube or iTunes U.* Topics like game theory and dynamical systems are standard topics for many university courses. YouTube, iTunes U, and similar sites have many videos of professors explaining them. Additionally, since these topics are so widely applicable, some of these explanations will be light on math and focus more on the concepts. These might be great complements to what we've done in this chapter.
- *Browse Ted.com for more on math's applications.* TED talks are short talks (typically 15–20 minutes) on various topics by experts in the field. There are many talks that discuss the applications of math. Hannah Fry, a British mathematician, has a short talk about some of the topics we've discussed in this chapter. Geoffrey West, a physicist, also has a TED talk where he discusses the common pattern that describes how wealth, crime, and even cities grow.

Epilogue

■■■■■■■■■■■■■■■

Congratulations! You've reached the end of this book! I think you deserve recognition. We've discussed so much in this book and have seen so many applications of mathematics that I'm proud of you for sticking through it all. The hard part is over. I won't surprise you with any more equations; there are no more mathematical concepts I'll share with you. I do, however, want to wrap the package I've given you in a nice bow.

You see, my broader goals for this book were to help you see how useful mathematics can be and to do that using concrete examples we all care about (like our health and finances). And by "mathematics" I really meant precalculus-level math. I set that as the math ceiling for this book because the topics covered in a precalculus class—algebra, geometry, functions, probability, and other miscellaneous topics—are things we spent several years studying in high school. And even if you didn't remember a thing about those topics, I was betting that I could gently remind you of the main points along the way.

Now that we're here, we get to take a bird's-eye view of the whole adventure. Here's the first thing I want to say: isn't it incredible how much we were able to do with just precalculus-level math? I mean, we discovered some powerful results, like the GS algorithm and the equation for the time until financial independence. That's one of the themes of this book: *mathematics is powerful*.

You may also have noticed how certain topics kept popping up in different situations. For instance, the linear equations you learned about in school show up, as we've now seen, in everything from your diet to your taxes to your happiness. Each time they do, they have something different to say, even though they're always the same old $y = mx + b$ equations. This illustrates another theme of the book: *mathematics is universal*; the same equations/results could apply to *many* different situations.

Even within the book's precalculus ceiling, we saw how more advanced math was able to describe more complicated phenomena (think of Chapters 5 and 6). That's an illustration of a general characteristic of mathematics: *the higher the level of mathematics used, the more powerful the results become.* (This is yet another good reason to study more math.) I'll give you another example from physics. Galileo used polynomials (covered in algebra) to describe motion on Earth—quite a big accomplishment for his time. Newton used calculus to describe motion in the entire universe via Newton's Laws of Motion (wow!). Einstein used multivariable calculus, differential equations, and other tools (like differential geometry) to one-up Newton and describe how gravity really works, predict black holes, describe how the universe started (and will end), and prove that time-travel is possible. *Holy cow*!

If you find these examples as awe-inspiring as I do, let me plead with you to speak up. I want you to share your positive experience with math. Not enough of us do that (many more of us share our negative experiences with math), and it shows in everything from our nation's international ranking in mathematics education to our individual fear of math. I want you to help me convince other people, especially young kids, that *math is worth learning.* More people need to know that mathematics is fun, that math is *genuinely useful,* and that math, contrary to what we might feel, *can be understood by anyone.*

I hope that you enjoyed the book. If you put together the themes I mentioned earlier you're led to one simple conclusion: *learn more math*! You're already off to a good start. You picked up a book about math. You read it. You thought about the math. Maybe not all of it made sense (if it did, please help others understand math!), but you stuck through it, and the mathematics led us to insights we wouldn't have achieved otherwise. There are so many more useful, fascinating, beautiful, and stunning applications and aspects of mathematics out there. All you've got to do now is keep up your math momentum. Here's to hoping that momentum never dies.

—Oscar E. Fernandez

Acknowledgments

■■■■■■■■■■■■■■■■

This book wouldn't have been possible without my editor at Princeton University Press, Vickie Kearn, and her team. Vickie's support and encouragement and her team's hard work on the project were incredibly helpful in preparing the book for publication. Without them, this book would still be a few scattered ideas in my head. I'm also grateful to the reviewers. They carefully read the first drafts of the book and gave me excellent feedback.

I also thank my wife, Zoraida. Part de facto copyeditor, part nonmath person test subject, Zoraida helped mold this book into something that (hopefully) makes sense to nonmathematicians. She also supported me throughout the long hours I spent putting thoughts to paper.

Finally, I want to thank *you*. I could've written the best book ever and it wouldn't have made a difference if nobody read it. The fact that *you* read it validates the hard work of all these people. Thank *you*.

Appendix A: Background Content

■■■■■■■■■■■■■■■■

In the simplest cases, to solve an algebraic equation we undo what's being done to the independent variable (usually x). For example, consider the equation

$$2x + 7 = 15.$$

(Note: a number next to variable—as in the case of $2x$—indicates multiplication.) Here x is being multiplied by 2, the result added to 7, and then set equal to 15. To solve for x we subtract 7 from both sides, leaving us with $2x = 8$, and then divide both sides by 2, giving $x = 4$.

There are several properties that can make algebraic manipulations easier. Here are the main ones we'll make use of.

- *The distributive law:* $a(b+c) = ab + ac$. Example: $3(x-2) = 3x - 6$.
- *The rules of exponents.* First, a few definitions. For any positive number a and any positive integer n (i.e., a number of the form 1, 2, 3, ...), a^n is defined to be a multiplied by itself n times. Example: $a^2 = a \times a$. Here are other properties we may run into (where applicable, b is another number like a, and m another like n).

  (1) $a^0 = 1$.
  (2) $a^{-n} = \frac{1}{a^n}$. Example: $2^{-2} = \frac{1}{4}$.
  (3) $a^{1/n} = \sqrt[n]{a}$. In the special case that $n = 2$ we write $a^{1/2} = \sqrt{a}$.
  (4) $a^m \times a^n = a^{m+n}$. Example: $2^3 \times 2^2 = 2^5$.
  (5) $\frac{a^m}{a^n} = a^{m-n}$. Example: $\frac{2^4}{2^2} = 2^2$.
  (6) $(ab)^n = a^n b^n$. Example: $6^5 = 2^5 \times 3^5$.
  (7) $\left(\frac{a}{b}\right)^n = \frac{a^n}{b^n}$. Example: $\left(\frac{2}{3}\right)^2 = \frac{4}{9}$.

FUNCTIONS

Sometimes we're interested in thinking of equations as "input-output machines." For example, consider the equation $y = x^3$. If we input an x-value (say, $x = 3$) the equation outputs a y-value that is the cube of the x-value (continuing, $y = 3^3 = 27$). Mathematicians tend to think about equations of the form "$y =$ stuff involving x's" this way. When dealing with such equations we say that "y is a *function of x*" if y depends on x, and each x-value is associated with exactly one y-value. For example, $y = x^3$ defines a function.[1]

Sometimes we also use the notation $f(x)$ (read "f of x") for functions; for example, $f(x) = x^2$. This notation is best understood from the "input-output machine" framework mentioned earlier. In this case we input an x-value into the function and it outputs an $f(x)$-value. For example, if $f(x) = x^2$ then $f(2) = 4$, and we'd say this out loud as "the value of the function when $x = 2$ is 4."

GLOSSARY OF MATHEMATICAL SYMBOLS

Here's a list of symbols used throughout this book along with an example of their usage.

Symbol	Meaning	Example of Usage
$=$	Equals; is the same as	$\dfrac{10}{2} = 5$
$\approx$	Is approximately	$1.001 \approx 1$
$\leq$	Less than or equal to; is at most	If $x \leq 2$, then $10x \leq 20$
$\geq$	Greater than or equal to; is at least	If $x \geq 3$, then $10x \geq 30$
$\pm$	Plus or minus	If $x = \pm 1$, then $x = 1$ or $x = -1$
$\implies$	Implies	$x = 4 \implies x^2 = 16$

---

[1] We clearly see that y depends on x, and that for a particular x-value there is only one y-value (and not more).

1. Pretend, for a moment, that x has a fixed value. Let's denote this fixed value by x_i ("i" for "initial"). Let's label the corresponding y-value y_i. The linear equation then reads

$$y_i = mx_i + b.$$

Now, let's add 1 to the x-value. We do this by replacing x_i in the equation by $x_i + 1$. Let's call the new y-value y_f ("f" for "final"). The new linear equation is

$$y_f = m(x_i + 1) + b.$$

We can distribute the m on the right-hand side of this equation: $m(x_i + 1) = mx_i + m$. This simplifies the equation to

$$y_f = mx_i + m + b.$$

Now look closely: the right-hand side is just m plus $mx_i + b$ (the order of the terms doesn't matter). But since $y_i = mx_i + b$, we can substitute this in to get

$$y_f = y_i + m.$$

Okay, let's recap what happened. After increasing the x-value by one unit (from x_i to $x_i + 1$), the new y-value (y_f) ended up being the initial y-value (y_i) plus the slope m. If m is positive, then y_f is bigger than y_i (the y-value has increased) whereas if it's negative, y_f is smaller than y_i (the y-value has decreased). That's the generalized slope interpretation I italicized on page 6.

2. Let's solve $4x + 370 \leq 400$ using algebra. (Since there aren't any negative numbers in our inequality the inequality sign gets treated the same as an equals sign.) Ready? Let's begin.

1. Here's the starting inequality: $\quad 4x + 370 \leq 400$
2. Now subtract 370 from both sides: $\quad 4x \leq 30$
3. Finally, divide both sides by 4: $\quad x \leq \frac{30}{4} = 7.5$

3. Here's how you would "mathematize" this problem. Let p be the total grams of protein eaten in a day, c the total grams of carbs, and f the total grams of fat. The Atwater general factor system tells us that the protein contains $4p$ calories, the carbs $4c$ calories, and the fat $9f$ calories. The total calories eaten, T, is then

$$T = 4p + 4c + 9f.$$

This equation is an example of a *multilinear* function. We'll discuss these in more detail in the next section. For now note that we can go through the same analysis of capping the total calories, T, to a certain number and then solving for the grams of each macronutrient. For example, capping T at 1,000 yields the inequality

$$4p + 4c + 9f \leq 1{,}000.$$

If we know two of three variables in this equation we can solve for the remaining variable. For example, if you wanted to stick to a diet low in carbs (say, $c = 150$) and fat (say, $f = 20$), then your diet would have at most 55 grams of protein (i.e., $p \leq 55$).

4. The full RMR_m equation involves four variables. To graph it would require a four-dimensional graph, which we can't visualize. But if I plug in a height, say, $h = 67$, we get an equation with three variables:

$$\text{RMR}_m = 4.5w - 5a + 1{,}070.3. \tag{A1.1}$$

This equation requires a three-dimensional graph. But that's okay; we graph in 3D just like we graph in 2D. We first draw the xy-plane on the

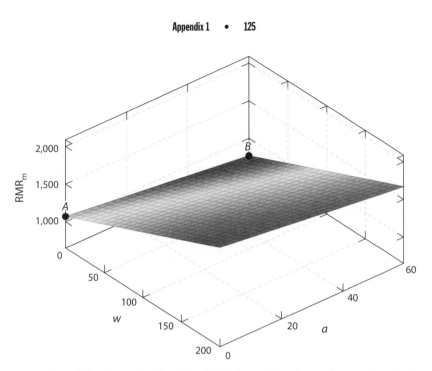

Figure A1.1. The 3D graph of equation (A1.1) for weight values w between 0 and 200 and age values a between 0 and 60.

bottom (like a floor) and then add a third axis going up. Then we plot a bunch of points relative to the origin (defined to be where the upward axis intersects the plane) and connect the dots. Figure A1.1 shows the graph of (A1.1) (called a *plane*). Planes are multilinear functions (note the lines that make up the edges of the plane in the figure). To illustrate that, notice that setting $w = 0$ gives $RMR_m = -5a + 1{,}070.3$. This is the downward sloping line (the slope is -5) connecting the points labeled A and B in the figure.

5. Starting from $20 = 0.15r - 8.85$, we ...

    a. Add 8.85 to both sides: $0.15r = 28.85,$

    b. Divide both sides by 0.15: $r = \frac{28.85}{0.15} = 192.3 \approx 192.$

Here $\approx$ means "approximately." (I've put a list of the mathematical symbols in the Glossary of Mathematical Symbols in Appendix A.)

TABLE A1.1.
The general forms of polynomials of degree 0 to 3, along with their names and specific examples.

n	Polynomial	Name	Example
0	a_0	Constant function	$y = 2$
1	$a_1 x + a_0$	Linear function	$y = 4x + 370$
2	$a_2 x^2 + a_1 x + a_0$	Quadratic function	$y = -0.007x^2 + 192$
3	$a_3 x^3 + a_2 x^2 + a_1 x + a_0$	Cubic function	$y = x^3 - 2x^2 + 1$

6. Every polynomial has the form

$$y = a_n x^n + a_{n-1} x^{n-1} + \ldots + a_1 x + a_0$$

for some numbers a_0, a_1, up to a_n (we assume $a_n \neq 0$), and some non-negative whole number n. The number n in this equation is called the *degree* of the polynomial; it's the highest power of x present. Table A1.1 gives the general form of polynomials of degree 0 to 3, along with their names and concrete examples.

7. To find the answer we set $\text{MHR} = \text{MHR}_{\text{pop}}$:

$$220 - a = 192 - 0.007a^2.$$

Adding $0.007a^2$ to both sides and subtracting 192 from both sides yields

$$0.007a^2 - a + 28 = 0.$$

The fastest way to solve this is to use the *quadratic formula*, which says that the solutions to $Ax^2 + Bx + C = 0$ are

$$x = \frac{-B \pm \sqrt{B^2 - 4AC}}{2A}.$$

The $\pm$ symbol means "plus or minus" (see the Glossary of Mathematical Symbols in Appendix A). It tells us to write down two solutions: one that uses the $+$ sign and another that uses the $-$ sign. Comparing $Ax^2 + Bx + C = 0$ to $0.007a^2 - a + 28 = 0$, we see that $A = 0.007$,

$B = -1$, and $C = 28$ (and $x = a$). The quadratic formula then gives the two solutions

$$a = \frac{20(25 - 3\sqrt{15})}{7} \approx 38.2, \qquad a = \frac{20(25 + 3\sqrt{15})}{7} \approx 104.6.$$

The first solution is the age (a-value) of the visible intersection point in Figure 1.2(b); the other solution corresponds to the other intersection point (not shown on the graph).

8. I'll show you how to mathematize this using Jason's ACB equation. Let t be the number of minutes it takes him to burn c calories. This means that

$$\text{ACB} = \frac{c}{t},$$

since ACB is the aerobic caloric burn *per minute*. This, together with Jason's ACB equation implies that

$$\frac{c}{t} = 0.15r - 8.85.$$

Since Jason's MHR is about 192 bpm, then $x\%$ of that is $\frac{192x}{100}$. (For example, to find 50% of his MHR we'd first divide 50 by 100 and then multiply the result by 192.) Thus, Jason will be exercising at this heart rate:

$$r = \frac{192x}{100}.$$

Inserting this into the previous equation yields

$$\frac{c}{t} = 0.15 \left(\frac{192x}{100} \right) - 8.85 \Rightarrow \frac{c}{t} = 0.228x - 8.85.$$

To solve for t we take the reciprocal of both sides (the reciprocal of $\frac{a}{b}$ is $\frac{b}{a}$) and then multiply both sides by c:

$$t = \frac{c}{0.288x - 8.85}.$$

For example, if Jason wanted to burn 400 calories ($c = 400$) by exercising at 70% ($x = 70$) of his MHR, this analysis estimates it would take him about $t \approx 46$ minutes.

■■■■■■■■■■■■■■■■

1. We can use the banana's energy density to calculate how many grams of bananas yield 100 calories:

$$\frac{100 \text{ calories}}{0.95 \ \frac{\text{calories}}{\text{gram}}} \approx 105 \text{ grams.}$$

Using a similar calculation we see that just $100/3.7 \approx 27$ grams of croissant yield 100 calories.

2. Let's look at a special case to keep things concrete. Pretend that $1/0$ had a definite value. Let's call that value x. Thus,

$$\frac{1}{0} = x.$$

By the rules of algebra we can multiply both sides by 0 to get

$$1 = x(0). \tag{A2.1}$$

But anything multiplied by zero gives zero. Therefore we get $1 = 0$. Clearly that's a problem.

Someone might then say "What about $0/0$? Following the same steps wouldn't lead to a problem." But it does, except a different kind of problem. In that case (A2.1) would become $0 = x(0)$. This is true for $x = 1$, $x = 2$, and actually *any value of x*. So, in this case you'd get an even more crazy conclusion: $0/0$ can have whatever value you'd like! To avoid these loony situations we mathematicians put our collective foot down and say "division by zero is not allowed" (by which we really mean "division by zero is not well defined").

3. Table 3 of [30] contains the data on YLL and WHtR. Putting these into a spreadsheet, graphing them as a scatterplot, and then having the program "add a trendline" gives the following equations for men's YLL:

$$y_{m,30} = 616.67r^3 - 920r^2 + 467.83r - 81,$$
$$y_{m,50} = 183.33r^3 - 180r^2 + 45.167r - 0.5, \qquad \text{(A2.2)}$$
$$y_{m,70} = -83.33r^3 + 245r^2 - 188.67r + 43.5,$$

and these for women's YLL:

$$y_{f,30} = 150r^3 - 175r^2 + 69r - 9.4,$$
$$y_{f,50} = 116.67r^3 - 130r^2 + 48.33r - 6.4, \qquad \text{(A2.3)}$$
$$y_{f,70} = 60r^2 - 58.4r + 14.21.$$

(In all but one instance cubic polynomials fit the data best.)

4. Let me illustrate how to use equations (A2.2)–(A2.3) to derive ones that work for any age by using myself as a test case. I'm 32 years old and male, so my particular YLL equation is somewhere between the first two in (A2.2). There's a 20-year range between those two equations, and being 32 means that I'm 2 years into that 20 year range, or 10% in. A reasonable estimate would weigh the $y_{m,30}$ much more (since I'm closer to 30 than I am to 50). One way to do this is to use the estimate

$$0.9y_{m,30} + 0.1y_{m,50}. \qquad \text{(A2.4)}$$

This quantity is 90% the value of $y_{m,30}$ plus 10% the value of $y_{m,50}$. To calculate it, I multiply the first line of (A2.2) by 0.9 and add the result to 0.1 times the second line of (A2.2). A similar approach can be used to generate YLL equations for any age beyond 30.

Appendix 3

■■■■■■■■■■■■■■■■

1. According to Table 3.1, the tax due in the 15% bracket is $922.50 plus 15% of the amount in excess of $9,225. Mathematically, the "amount in excess of $9,225" is $z - 9,225$. To find 15% of that we multiply by 0.15. Finally, adding $922.50 produces (3.3).

2. First, a quick refresher on working with percentages. As an example, if a gallon of gasoline currently costs $3 and the price increases by 50%, then the new price is $3 plus 50% of $3:

$$\$3 + (\$3) \left(\frac{50}{100} \right) = \$4.50.$$

Notice that we can factor out the $3 from the left-hand side of this equation, so that it becomes:

$$\$3 \left(1 + \frac{50}{100} \right) = \$4.50.$$

So we've learned that:

$$(\text{Current price}) \left(1 + \frac{\text{Percent increase}}{100} \right) = \text{New price.}$$

Let's now apply this knowledge to the 19¢ cheeseburger. Let's call x the unknown percentage increase. Then

$$19 \left(1 + \frac{x}{100} \right) = 100.$$

Simplifying and solving for x gives:

$$1 + \frac{x}{100} = \frac{100}{19} \implies x = 100\left(\frac{100}{19} - 1\right) \approx 426.31.$$

3. The equation

$$19\left(1 + x\right)^{60} = 100$$

might look intimidating at first. But since our goal is to solve for x, we want to undo everything that's being done to x, so let's talk through that. First, we add 1 to x. Then we raise that number to the 60th power. Finally, we multiply by 19 and set the result equal to 100. So all we have to do is go in reverse: divide 100 by 19, raise the result to the $(1/60)$th power, and then subtract 1. Here's that in math form:

a. Here's the starting equation: $\qquad\qquad 19(1 + x)^{60} = 100$

b. Now divide both sides by 19: $\qquad\qquad (1 + x)^{60} = \frac{100}{19}$

c. Next, raise both sides to the $(1/60)$th power: $\quad 1 + x = \left(\frac{100}{19}\right)^{\frac{1}{60}}$

Finally, subtract 1 to get

$$x = \left(\frac{100}{19}\right)^{\frac{1}{60}} - 1 \approx 0.0281.$$

This is the decimal form of the percentage 2.81%.

4. We perform a calculation similar to the one in point 3 above:

$$27\left(1 + x\right)^{70} = 855 \implies x = \left(\frac{855}{27}\right)^{\frac{1}{70}} - 1 \approx 0.0506,$$

which is decimal form for 5.06%.

5. For simplicity, let's study the exponential function $y = 2^x$. Now, pick your favorite x-value; let's label that number x_0. The correspoding y-value (let's call it y_{initial}) is

$$y_{\text{initial}} = 2^{x_0}.$$

A one-unit change in x_0 yields $x_0 + 1$, and its corresponding y-value is

$$y_{final} = 2^{x_0+1}.$$

But by the rules of exponents (see Appendix A for a refresher),

$$2^{x_0+1} = 2^{x_0}2^1,$$

so that

$$y_{final} = 2^{x_0}2^1 = 2y_{initial}.$$

In other words, a one-unit increase in x caused the y-value to be multiplied by 2, which is the base of the exponential function $y = 2^x$. Were we to go back and replace $y = 2^x$ with $y = ab^x$, a similar calculation would yield the interpretation for b found in the chapter.

6. Let's compare the linear function $y = 2x$ with the exponential function $y = 2^x$. Notice that the only difference is in the placement of the variable x. However, because the slope of a linear function and the base of an exponential function have two different interpretations, the placement of x makes all the difference. To see this take a look at Figure A3.1(a). Notice how every time we increase the x-value by 1 we *add* 2 to the y-value of the linear function (since $y = 2x$ has slope 2), whereas we *multiply* the y-value of the exponential function by 2 (since $y = 2^x$ has base 2). Doing this over and over produces the very different graphs in Figure A3.1(b).

7. Take the doubling of the 1¢! The exponential function describing the doubling process is

$$y = (0.01)2^x,$$

since 1¢ = $0.01 is the *initial* value, and since that gets *doubled* every day (here x represents the days since the doubling began). On day 30 you'll have ... drumroll please ...

$$(0.01)2^{30} = \$5{,}368{,}709.12!$$

Please send me a "consultant's fee" for helping you not choose the $1 million.

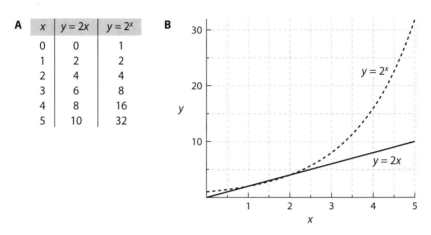

Figure A3.1. (a) A table of some values of the linear function $y = 2x$ and the exponential function $y = 2^x$. (b) The graphs of $y = 2x$ (solid line) and $y = 2^x$ (dashed curve).

8. Let's derive this formula by calculating the balance of the loan after each of the first two payments, and then spotting the pattern in the equations.

When your first statement arrives you're charged Lc in interest, the loan amount multiplied by the monthly interest rate (expressed as a decimal). This amount gets added to the principal (L). But then your monthly payment of M dollars reduces that sum. So the loan's balance after your first payment is

$$B_1 = L + Lc - M = L(1 + c) - M.$$

The same thing happens after the second payment; your loan balance then is

$$B_2 = B_1 + B_1 c - M$$

$$= L(1 + c) - M + [L(1 + c) - M]c - M$$

$$= L(1 + c) + Lc(1 + c) - M(1 + c) - M$$

$$= L(1 + c)^2 - M[(1 + c) + 1].$$

By comparing B_1 and B_2 we can predict what B_3 will be:

$$B_3 = L(1+c)^3 - M[(1+c)^2 + (1+c) + 1]. \qquad (A3.1)$$

(You can verify this by following the same steps in the B_2 calculation.) Notice that the term with the L always multiplies $(1+c)$ raised to the same power as the subscript of B. Also, notice that M multiplies an expression which is the sum of powers of $(1+c)$. In fact, we can simplify the expression in the brackets in the B_3 equation as follows. First, let's call it T_3:

$$T_3 = (1+c)^2 + (1+c) + 1.$$

Now here's a trick: let's multiply T_3 by $(1+c)$ and compare that to $-T_3$:

$$(1+c)T_3 = (1+c)^3 + (1+c)^2 + (1+c)$$
$$-T_3 = -(1+c)^2 - (1+c) - 1.$$

Notice that if we add these two equations we get

$$(1+c)T_3 - T_3 = (1+c)^3 - 1 \implies T_3 = \frac{(1+c)^3 - 1}{c}.$$

(You might have already learned this trick if you've studied geometric series.) Using this in (A3.1) we get

$$B_3 = L(1+c)^3 - M\left[\frac{(1+c)^3 - 1}{c}\right].$$

So after the nth payment the loan balance will be

$$B_n = L(1+c)^n - M\left[\frac{(1+c)^n - 1}{c}\right].$$

Since we've assumed that the loan is paid off after these n payments, we set $B_n = 0$. This yields

$$L(1+c)^n = M \left[\frac{(1+c)^n - 1}{c} \right] \implies M = \frac{Lc(1+c)^n}{(1+c)^n - 1}.$$

Finally, dividing the numerator and denominator of this equation by $(1+c)^n$ yields (3.10).

9. Here $L = 100,000$ and $r = 6$ (which makes $c = 6/1,200 = 0.005$), so:

$$M = \frac{(100,000)(0.005)}{(1 - (1 + 0.005))^{-360}} \approx 536.82.$$

10. The biggest extra expenses associated with home ownership are property taxes, private mortgage insurance—required if your down payment in less than 20% of the home's price—and homeowner's insurance. These can all be expressed as a percentage of the home's value. Let's annualize these expenses, express each as a percentage of the home's value, and call the sum of these percentages y (expressed in decimal form). Then, your monthly rent payment P now needs to pay both the mortgage—whose monthly payment is given by (3.10)—and these extra monthly costs, which are $yL/12$:

$$P = \frac{Lc}{1 - (1+c)^{-n}} + \frac{yL}{12} = L \left[\frac{c}{1 - (1+c)^{-n}} + \frac{y}{12} \right]. \quad \text{(A3.2)}$$

Now, in the chapter we neglected the extra costs we've now included, and the rent payment P was used solely to pay the mortgage. Let's call the cost of the original house you could afford L_{orig}. From (3.11) we know that

$$L_{\text{orig}} = \frac{P\left(1 - (1+c)^{-n}\right)}{c} \implies P = \frac{L_{\text{orig}}c}{1 - (1+c)^{-n}}.$$

Substituting this into the left-hand side of (A3.2) yields

$$\frac{L_{\text{orig}}c}{1-(1+c)^{-n}} = L\left[\frac{c}{1-(1+c)^{-n}} + \frac{y}{12}\right].$$

Solving for L (and simplifying) gives

$$L = \frac{L_{\text{orig}}}{1 + y\left(\frac{1-(1+c)^{-360}}{12c}\right)}. \tag{A3.3}$$

This new formula says that the extra expenses associated with owning a home *reduce* how much house you can afford (since we're dividing L_{orig} by a number greater than 1). As an example, let's consider converting a \$1,000 monthly rent payment to a 4% mortgage. Neglecting extra expenses, we get a loan amount of $L_{\text{orig}} \approx \$209,461$ (using (3.11)). Now suppose that the extra annual expenses amount to 2% of the home's value. Then, $y = 0.02$ and (A3.3) gives $L \approx \$155,260$, or about a 26% reduction in the house you can afford.

One final note. Though I've assumed that y is the sum of just three expenses, it could be the sum of *all* the expenses associated with owning a home that can be expressed as a percentage of the home's value (including, for example, landscaping).

11. Here are the steps:

a. Here's the starting equation: $\qquad M = \dfrac{Lc}{1-(1+c)^{-n}}$

b. Multiply both sides by $1-(1+c)^{-n}$: $\quad M\left[1-(1+c)^{-n}\right] = Lc$

c. Distribute the M: $\qquad M - M(1+c)^{-n} = Lc$

d. Isolate the term with n: $\qquad M - Lc = M(1+c)^{-n}$

e. Divide by M: $\qquad \frac{M-Lc}{M} = (1+c)^{-n}$

f. Finally, take the reciprocal of both sides: $\quad \frac{M}{M-Lc} = (1+c)^{n}.$

12. To explain the equivalence (3.13) we first take the base b logarithm of both sides of the equation $y = b^x$:

$$\log_b y = \log_b(b^x).$$

We now use two properties of logarithms: (1) $\log_b b^x = x \log_b b$, and (2) $\log_b b = 1$. We get:

$$\log_b y = x.$$

In other words, the "base b" logarithm helps solve $y = b^x$ for x. To bring things back to $\log x$ we can use the *change of base* formula:

$$\log_b y = \frac{\log_a y}{\log_a b}.$$

If we let $a = 10$, then we get

$$\log_b y = \frac{\log_{10} y}{\log_{10} b} = \frac{\log y}{\log b},$$

which matches the right-most equation in (3.13).

13. Equation (3.12),

$$(1+c)^n = \frac{M}{M - Lc},$$

can be written in the form

$$y = b^n, \quad \text{where} \quad y = \frac{M}{M - Lc} \quad \text{and} \quad b = 1 + c.$$

We then apply the equivalence (3.13), substituting in the equations for y and b from above. The result is (3.14).

14. My two favorite online graphing utilities are wolframalpha.com and desmos.com. On the WolframAlpha site, input

plot $y = \log(x/(x - 10))/\log(1.01)$ from $x = 20$ to $x = 100$

exactly as I've written it and you'll get the graph in Figure 3.3(a). On demos.com, as you start typing the function, the site will adjust the math accordingly. For example, type "log(" and another ")" will appear.

Once you finish typing in your function on desmos.com, you may need to zoom out to see the graph. One advantage of this site is that it's interactive—in addition to zooming and panning, you can click anywhere on the curve and it'll tell you the coordinates of that point.

15. The simplest explanation uses the following observation: when you make more than the minimum monthly payment you avoid paying interest on that money. For example, paying an extra $100 this month to a credit card that carries a 1% monthly interest rate (i.e., a 12% annual interest rate) reduces your next month's balance by $101. That's not a typo—next month's balance is reduced by the $100 you paid *and* by the interest the credit card company *doesn't get to charge you on that* $100 (that interest is 1% of $100, which is where the extra $1 in savings comes from). Therefore, to minimize the amount of interest paid you should apply extra payments to the highest-rate credit card.

16. After t years of saving S dollars per year and earning $r\%$ each year on the accumulated savings, that sum of money would be worth

$$S\left(\frac{(1+r)^t - 1}{r}\right), \qquad \text{where } r \text{ is in decimal form.} \qquad \text{(A3.4)}$$

Now we're also assuming that you currently have B dollars in savings. Using what we learned in Section 3.2.1 we know that after t years of earning $r\%$ on those B dollars in savings, that sum of money would be worth

$$B(1+r)^t, \qquad \text{where } r \text{ is in decimal form.} \qquad \text{(A3.5)}$$

Thus, in t years your total nest egg would be worth

$$N_t = S\left(\frac{(1+r)^t - 1}{r}\right) + B(1+r)^t.$$

Now all that's left is to solve $T = 0.04 N_t$, or

$$T = 0.04\left[S\left(\frac{(1+r)^t - 1}{r}\right) + B(1+r)^t\right].$$

Here are the steps:

a. The starting equation: $T = 0.04 \left[S \left(\dfrac{(1+r)^t - 1}{r} \right) + B(1+r)^t \right]$.

b. Divide both sides by 0.04: $\quad 25T = S\left(\dfrac{(1+r)^t - 1}{r} \right) + B(1+r)^t$

c. Multiply both sides by r: $\quad 25Tr = S((1+r)^t - 1) + Br(1+r)^t$

d. Distribute the S: $\qquad\qquad\quad 25Tr = S(1+r)^t - S + Br(1+r)^t$

e. Factor out $(1+r)^t$: $\qquad\qquad 25Tr = (S + Br)(1+r)^t - S$

f. Add S to both sides: $\qquad\quad\; 25Tr + S = (S + Br)(1+r)^t$

g. Divide both sides by $S + Br$: $\qquad \dfrac{25Tr + S}{S + Br} = (1+r)^t$.

The equivalence (3.13) then yields

$$t = \frac{\log \left(\dfrac{25Tr + S}{S + Br} \right)}{\log (1 + r)}. \qquad\qquad \text{(A3.6)}$$

Let's now focus on the term inside the parentheses in the numerator. If we divide the numerator and denominator by T we get:

$$\frac{\frac{25Tr + S}{T}}{\frac{S + Br}{T}} = \frac{25r + \frac{S}{T}}{\frac{S}{T} + \frac{Br}{T}} = \frac{25r + STE}{STE + \frac{Br}{T}}.$$

Replacing the term inside the parentheses in the numerator of (A3.6) with this new fraction yields (3.17).

17. Since (3.16) represents your total expenses, your gross income G can be expressed as

$$G = T + S,$$

the sum of your total expenses and your savings. Dividing both sides by G yields

$$1 = \frac{T}{G} + \frac{S}{G}. \qquad\qquad \text{(A3.7)}$$

Here, S/G is your *savings percentage*—the percentage of your gross income you save each year—and T/G your *expenses percentage*.

Now, I introduced the STE ratio by the formula $STE = S/T$. If we divide the numerator and denominator of this fraction by G we get

$$STE = \frac{\frac{S}{G}}{\frac{T}{G}} = \frac{\frac{S}{G}}{1 - \frac{S}{G}}, \qquad (A3.8)$$

where I used (A3.7) to get the last equation. For example, saving 20% of your gross income would yield an STE ratio of

$$STE = \frac{0.2}{1 - 0.2} = 0.25.$$

We can also use (A3.8) to express the savings percentage in terms of the STE ratio. Rearranging,

$$STE \left(1 - \frac{S}{G} \right) = \frac{S}{G} \implies \frac{S}{G} = \frac{STE}{1 + STE}. \qquad (A3.9)$$

For example, if your STE ratio is 2 (i.e., you save twice as much as you spend), then (A3.9) says that your savings percentage is

$$\frac{S}{G} = \frac{2}{1 + 2} = \frac{2}{3} = 66\%.$$

■■■■■■■■■■■■■■■■

1. This rule of thumb is based on the observation that you should only invest if your after-tax return is more than the money you'd save in interest by paying down debt. Since your investment gains will likely be taxed at 15%, to account for that your return needs to be about 18% *more* than the debt's annual interest rate:

$$\text{Investment return} \geq 1.18(\text{Debt's annual interest rate}).$$

(I rounded off the 1.18 to 1.2 in the chapter.)

2. Here they are.

- *IPO*: Initial public offering. This document contains the details of the offering, including how many shares of stock the company will make available to the general public, what *exchange* (e.g., the New York Stock Exchange) the stock will be listed on, what the initial price of the shares would be, and when they would start trading.
- *Dividend*: Periodic payments to investors of the company's profits.
- *Corporate bond*: Debt issued by a corporation. An investor who buys a corporate bond is loaning a corporation money until a specific date (called the *maturity date*), at which point they'll receive back their initial investment (assuming the company doesn't default on the bond; see the next definition). The investor also receives interest payments every year (or quarter, or month) from the corporate bond up until the maturity date.
- *Default*: When a corporation—or other entity that issues debt, like a government or a city—cannot pay the full amount due to its bondholders.

- *Downgrade*: Rating agencies rate a bond according to its issuer's default risk. A downgrade indicates that the default risk his risen.
- *Brokerage firm*: A financial institution that buys/sells securities on behalf of an investor. Note: many brokerages charge a *commission* (fee) for each trade.
- *S&P 500*: A stock market index founded in 1957 that includes 500 of the largest publicly traded companies in the United States.
- *Treasury bonds*: Bonds issued by the U.S. Department of Treasury (otherwise known as "government bonds").

Appendix 5

1. The site lists Boston's population (as of 2013) as $P = 645,966$, with about 52% of that being women (so $S = 0.52$). I'll keep things simple and use 30–40 as my age preference. Using another table on the site I calculated that of the roughly 335,902 women in Boston, about 15% are between 30 and 40 (so $A = 0.15$). The same site says that 23.4% of Bostonians age 25 or older had at least a bachelor's degree ($E = 0.234$) and that 54.2% aren't married; I'll assume that only half of those are up for dating (so that $D = 0.271$). Finally, I'll assume that I'd only consider dating 1/3 of the remaining women ($H_1 = 0.33$) and that 1/3 of them would consider dating me ($H_2 = 0.33$). That gives

$$N = (645,966)(0.52)(0.15)(0.234)(0.271)(0.33)(0.33) \approx 350.$$

2. Here's the proof that everyone gets matched up. If we're wrong then there remain at least two lonely people—let's call them Larry and Laura—who aren't engaged. That means Laura never received a proposal. But that's impossible, because Laura is somewhere on Larry's list. According to Mary's rules, Larry had to continue proposing to women he hadn't already proposed to, which would eventually lead him to propose to Laura. Another one of Mary's rules—that you must accept the proposal if it's the only one you receive—would then guarantee that Laura would have accepted the proposal.

3. The men end up with their best partners because they do the proposing in order from best to worst. The women, on the other hand, only get to trade up. So they're effectively choosing in order from worst to best.

APPENDIX 6

■■■■■■■■■■■■■■■■

1. Recall from Chapter 3 that $y = ab^x$ (where $b > 0$ and $b \neq 1$) is the general form of an exponential function. Every such function can be rewritten in the form $y = ae^{rx}$, where r is called the *continuous growth rate* (if $r > 0$) or *decay rate* (if $r < 0$) and $e \approx 2.71$ is Euler's number (which I mentioned back in Section 5.2).[1] To figure out the instantaneous change in y, let me give you a very short introduction to derivatives, one of the two pillars of calculus.

The phrase "instantaneous change" is, mathematically, problematic. How can something *change* in an *instant*? So, what mathematicians have decided to do is to first derive a formula that describes how something changes over a *very small* interval (we call this the *average rate of change*), and then see if we can shrink that interval to zero in that formula. If nothing goes wrong, we call the result the *derivative* and use that as our instantaneous change. Here's how this process works in our context.

To start, let's put a subscript on y that tracks the x-value, so that $y_x = ae^{rx}$. Let's now see what happens to y_x when we increase the x-value by h (think of h as being a positive and *very small* number, like 0.000001):

$$y_{x+h} = ae^{r(x+h)} = ae^{rx+rh} = ae^{rx}e^{rh}.$$

Note that we've increased the x-value by h and that's changed the y-value from y_x to y_{x+h}. Hey, that sounds like our slope interpretation for the slope of a line! (Recall the discussion in the first few pages of

---

[1]To see how, note that the two equations are the same provided $b = e^r$. Using (3.13), this implies that $r = \frac{\log b}{\log e}$.

Chapter 1.) You might remember how we calculate slopes of lines—take the ratio of the change in y to the change in x. In our case:

$$\frac{\text{Change in } y}{\text{Change in } x} = \frac{y_{x+h} - y_x}{(x+h) - x} = \frac{ae^{rx}e^{rh} - ae^{rx}}{(x+h) - x}.$$

Simplifying the second and third equations yields

$$\frac{y_{x+h} - y_x}{h} = ae^{rx}\left(\frac{e^{rh} - 1}{h}\right). \qquad (A6.1)$$

The quantity on the left-hand side is the average rate of change of the function $y = ae^{rx}$ over the interval x to $x + h$.

Okay now comes the fun part: does anything go wrong on the right-hand side of (A6.1) when we make h as close to zero as we'd like? If the answer is "no" we replace the left-hand side of (A6.1) with y'_x (the notation for the derivative of y_x) and the right-hand side with whatever we calculate the "nearly $h = 0$" expression to be. Unfortunately, that calculation involves more calculus than I have space for here (specifically, something called *limits*). Instead, what I'll do is give you a rough idea of what is done.

First, to get at the instantaneous change we need to make h as small as possible. We can't make h zero because h is in the denominator and we can't divide by zero. So let's "phone a friend," in this case wolframalpha.com; if you pull up that site and type this in:

$$\text{plot y} = (e^{\wedge}(5h) - 1)/h \text{ from y} = 0 \text{ to } 10, \qquad (A6.2)$$

you'll see a graph which has a y-value of 5 when h is very close to 0. Change the 5 in (A6.2) to any other number between 0 and 10 and you'll notice the same thing: whatever number you chose is the y-value when h is very close to 0. Now look back at the fraction in the parentheses in (A6.1). Notice that the 5 in (A6.2) is the r in (A6.1). Thus, we have reason to believe that:

$$\text{When } h \text{ is very close to zero } \frac{e^{rh} - 1}{h} \approx r.$$

So the right-hand side of (A6.1) is approximately $ae^{rx}(r)$. But since ae^{rx} is just y_x, this simplifies even further to ry_x. Nothing bad happened now that we made h very close to zero, so we deduce that

$$y'_x = ry_x.$$

The takeaway: whenever a quantity is described by an exponential function (here that was y_x), its instantaneous change is equal to the continuous growth (or decay) rate of that exponential function (r) multiplied by the y-value of the function (y_x).

2. Since the equilibrium states are the states when feelings aren't changing, to find them we set x' and y'—the objects that represent the change in feelings—to zero in (6.3a)–(6.3b). Doing so yields the two equations

$$y = \frac{g(A_y) + R_x(y)}{d_y}, \qquad x = \frac{f(A_x) + R_y(x)}{d_x}.$$

In [55] the authors use realistic assumptions on the functions f, g, R_x, and R_y to show that these equations are satisfied at either one point, or three points. In the former that point has both x and y positive (i.e., the robust couple case). In the latter, two of the three points have both x and y negative, while one has both x and y positive; this case corresponds to the fragile couple.

3. In [55] the authors extend (6.3) to the setting of $2N$ couples each composed of one man and one woman. Assuming that no two women (or men) have the same appeal, they show that the matchings are stable only when the partner of the nth most appealing woman is the nth most appealing man. (Again, though, this is appeal relative to the other members of the same gender.) You might wonder if this solution to the stable matching problem has any relationship to the GS algorithm (Section 5.3). I don't know if there's a mathematical connection between the two (this is how research questions come up in math), but it makes intuitive sense that these "equal relative appeal couples" would be stable.

4. Notice that because x and z range from 0 to 500, $Y(x)$ ranges between 0 and 10 and $P(z)$ between 0 and 8. This verifies the first assumption. The second assumption follows from the fact that the slope of $Y(x)$ is greater than that of $P(z)$. Also, both functions have y-values of 0 when $x = 0$ or $z = 0$ (so that the third assumption holds). Finally, if $x = 500$ then $Y = 10$, whereas if $z = 500$ then $P = 8$ (this verifies the last assumption).

5. From (6.4) and (6.5) we can write $P(z)$ in terms of x:

$$P(x) = \frac{8}{500}(500 - x) = 8 - \frac{8}{500}x.$$

Then, since $x = \frac{500Y}{10}$ (from (6.4)), we can relate P to Y:

$$P = 8 - \frac{8}{500}\frac{500Y}{10} = 8 - \frac{8}{10}Y.$$

Let's call the product $(Y - 3)(P - 4)$ the variable N (for Nash). Using the relationship we just derived, we get

$$N = (Y - 3)(P - 4)$$
$$= (Y - 3)\left(8 - \frac{8}{10}Y - 4\right)$$
$$= (Y - 3)(4 - 0.8Y) \qquad\qquad (A6.3)$$
$$= -0.8Y^2 + 6.4Y - 12.$$

Since the coefficient of the quadratic term is negative (-0.8), the graph of N is a parabola that opens downward (see Figure A6.1). That's good news, because it means there's a maximum. That max is located halfway between the *zeros* of the function (the Y-values that make N zero). From (A6.3) we see that $N = 0$ when $Y = 3$ or $Y = 4/0.8 = 5$. (You can also see this from Figure A6.1.) Thus, the maximum occurs at $Y = 4$ (the midpoint of 3 and 5). Finally, from (6.4) and (6.5) this gives the x- and z-values:

$$x = 200, \qquad z = 300.$$

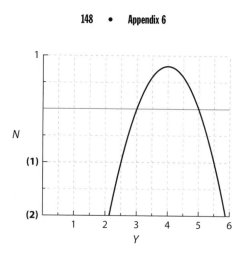

Figure A6.1. The graph of the quadratic function $N = -0.8Y^2 + 6.4Y - 12$.

6. We'll follow the same process we did for dividing up $500, except now we'll be dividing up a total T of some quantity. You get x of that, and your partner gets z, so the constraint is

$$x + z = T. \qquad (A6.4)$$

Using again the 0 to 10 scale for measuring happiness, and again assuming linear utility functions, we get

$$Y(x) = \frac{M}{T}x, \qquad P(z) = \frac{N}{T}z. \qquad (A6.5)$$

As a quick check, if you get all of T (i.e., $x = T$), then the first equation says that $Y = M$. Thus, M is the maximum happiness you'd experience. Similarly, N is the maximum happiness your partner would experience.

Now, solving (A6.4) for z and plugging that into the equation for $P(z)$ in (A6.5) gives

$$P(x) = \frac{N}{T}(T - x) = N - \frac{N}{T}x. \qquad (A6.6)$$

Solving the $Y(x)$ equation in (A6.5) for x gives $x = \frac{TY}{M}$, and using this in (A6.6) gives

$$P = N - \frac{N}{T}\frac{TY}{M} = N - \frac{N}{M}Y. \qquad \text{(A6.7)}$$

(This tells us that you and your partner's utility functions are related.)

Let's move on to the Nash product. Let's call it H this time (since N is already taken):

$$H = (Y - Y_d)(P - P_d).$$

Inserting (A6.7) gives

$$H = (Y - Y_d)\left(N - \frac{N}{M}Y - P_d\right). \qquad \text{(A6.8)}$$

Were we to multiply this out, we'd get a quadratic equation in Y. Moreover, the coefficient of the quadratic term is $-N/M$, which, since it's negative, indicates that the graph of (A6.8) is a downward-opening parabola (similar to Figure A6.1). Following our procedure for finding the maximum from before, we first find the zeros of H. They are:

$$Y = Y_d, \qquad N - \frac{N}{M}Y - P_d = 0.$$

Solving the second equation yields

$$Y = \frac{M}{N}(N - P_d) = M - \frac{M}{N}P_d.$$

Finally, we take the average of this and $Y = Y_d$ to get the location of the maximum:

$$Y_{\max} = \frac{1}{2}\left(Y_d + M - \frac{M}{N}P_d\right).$$

Factoring out an M and rearranging yields

$$Y_{\max} = \frac{M}{2}\left(1 + \frac{Y_d}{M} - \frac{P_d}{N}\right). \tag{A6.9}$$

This is your utility, but your share of T is found by using the prior relationship of $x = \frac{TY}{M}$. Substituting in (A6.9) here—which amounts to multiplying (A6.9) by T/M—gives (6.7a). Then, substituting the result in $z = T - x$ gives (6.7b).

7. If $Y_d = P_d$ but they're not zero, then (6.7b) simplifies to

$$z = \frac{T}{2}\left(1 + P_d\left(\frac{1}{N} - \frac{1}{M}\right)\right) = \frac{T}{2}\left(1 + P_d\left(\frac{M - N}{MN}\right)\right).$$

Distributing the $T/2$ yields

$$z = \frac{T}{2} + (M - N)\frac{P_d T}{2MN}.$$

Since we've assumed that $N > M$, the $(M - N)$ part of the second term is negative. This means that z is equal to $T/2$ *minus* a positive number. In other words, your partner gets *less than* half of T, meaning that they get a smaller share of T than you (who gets more than half of T).

8. When $M = N$ (6.7a) simplifies to

$$x = \frac{T}{2}\left(1 + \frac{1}{M}(Y_d - P_d)\right).$$

As before, let's distribute the $T/2$:

$$x = \frac{T}{2} + (Y_d - P_d)\frac{T}{2M}.$$

If $Y_d > P_d$ then the second term is positive, meaning that x is larger than $T/2$. Thus, you get more than half of T. However, if $P_d > Y_d$ the second term is *negative*. In this case you get *less than* half of T.

Bibliography

■ ■ ■ ■ ■ ■ ■ ■ ■ ■ ■ ■ ■ ■ ■ ■ ■

I highly encourage you to read through the references I've cited. To make it easier I've indicated which ones can be read for free online. You may be able to read through the others at a public library or a university library, or you can search the web—sometimes authors upload their full-text publications to their website. I've also added some comments regarding the limitations of the results reported in some of the references.

[1] *Food energy—Methods of Analysis and Conversion Factors.* Report of a Technical Workshop. FAO Food and Nutrition Paper 77. Rome: Food and Agriculture Organization (2003). (Free to read.)

[2] Dalton, S. *Overweight and Weight Management: The Health Professional's Guide to Understanding and Treatment.* Gaithersburg, MD: Aspen Publishers (1997).

[3] Frankenfield, D., et al., "Comparison of Predictive Equations for Resting Metabolic Rate in Healthy Nonobese and Obese Adults: A Systematic Review," *Journal of the Academy of Nutrition and Dietetics*, 105, no. 5 (2005), 775–789. (Free to read.)

    Comments. The authors note that "the Mifflin–St. Jeor equation is more likely than the other equations tested to estimate RMR to within 10% of that measured, but noteworthy errors and limitations exist when it is applied to individuals and possibly when it is generalized to certain age and ethnic groups."

[4] Willis, L. H., et al., "Effects of Aerobic and/or Resistance Training on Body Mass and Fat Mass in Overweight or Obese Adults," *Journal of Applied Physiology*, 113, no. 12 (2012), 1831–1837. (Free to read.)

[5] Keytel, L. R., et al., "Prediction of Energy Expenditure from Heart Rate Monitoring during Submaximal Exercise," *Journal of Sports Science*, 23 (2005), 289–297.

    Comments. The authors note that (1.4) accounts for "73.4% of the variance in energy expenditure in this sample," so the formula isn't perfect. Moreover, the heart rates of the participants in the study ranged between 100 bpm and 180 bpm, so (1.4) shouldn't be trusted to estimate ACB outside that heart rate range.

[6] Gellish, R. L., et al., "Longitudinal Modeling of the Relationship between Age and Maximal Heart Rate," *Medicine & Science in Sports & Exercise*, 39, no. 5 (2007), 822–829.

[7] Halton, T. L., and Frank, B. H. "The Effects of High Protein Diets on Thermogenesis, Satiety and Weight Loss: A Critical Review," *Journal of the American College of Nutrition*, 23, no. 5 (2004), 373–385.

[8] Scott, C. B., et al., "Onset of the Thermic Effect of Feeding (TEF): A Randomized Cross-over Trial," *Journal of the International Society of Sports Nutrition*, 4, no. 24 (2007). (Free to read.)

[9] Helms, E. R., et al., "Evidence-Based Recommendations for Natural Body-building Contest Preparation: Nutrition and Supplementation," *Journal of the International Society of Sports Nutrition*, 11, no. 20 (2014). (Free to read.)

[10] "Triglycerides: Why Do They Matter?," *Diseases and Conditions: High Cholesterol*, Mayo Foundation for Medical Education and Research, n.d. http://www.mayoclinic.org/diseases-conditions/high-blood-cholesterol/in-depth/triglycerides/art-20048186 (Accessed 15 May 2015). (Free to read.)

[11] Phillips, S. M. "Dietary Protein for Athletes: From Requirements to Metabolic Advantage," *Applied Physiology, Nutrition, and Metabolism*, 31 (2006), 647–654.

[12] Bellisle, F., et al., "Meal Frequency and Energy Balance," *British Journal of Nutrition*, 77, no. S1 (1997), S57–S70.

[13] Ohkawara, K., et al., "Effects of Increased Meal Frequency on Fat Oxidation and Perceived Hunger," *Obesity*, 21, no. 2 (2013), 336–343. (Free to read.)

[14] La Bounty, P. M., et al., "International Society of Sports Nutrition Position Stand: Meal Frequency," *Journal of the International Society of Sports Nutrition*, 8, no. 4 (2011). (Free to read.)

[15] Gunnars, K. "23 Studies on Low-Carb and Low-Fat Diets—Time to Retire the Fad," *Authoritynutrition.com*, n.d. http://authoritynutrition.com/23-studies-on-low-carb-and-low-fat-diets/ (Accessed 15 May 2015). (Free to read.)

[16] Westman, E. C., et al., "Low-Carbohydrate Nutrition and Metabolism," *American Journal of Clinical Nutrition*, 86, no. 2 (2007), 276–284. (Free to read.)

[17] *Fats and Oils in Human Nutrition*, Report of a Joint FAO/WHO Expert Consultation, FAO Food and Nutrition Paper 57. Rome: Food and Agriculture Organization (1993). http://www.fao.org/docrep/v4700e/V4700E08.htm (Free to read.)

[18] Becker, K. L. *Principles and Practice of Endocrinology and Metabolism*, Philadelphia, PA: Lippincott, Williams and Wilkins (2001).

[19] Parks, E. J. "Effect of Dietary Carbohydrate on Triglyceride Metabolism in Humans," *Journal of Nutrition*, 131, no. 10 (2001), 2772S–2774S. (Free to read.)

[20] Mensink, R. P., et al., "Effects of Dietary Fatty Acids and Carbs on Blood Lipids," *American Journal of Clinical Nutrition*, 77 (2003), 1146–1155. (Free to read.)

Comments. The authors cite several limitations of their study, including the fact that "the studies included in our meta-analysis lasted between 13 and 91 d [days]." As they state, "this raises the question of whether the effects observed are transitory." See also the discussion in Section 2.1.4 regarding the error ranges of the coefficients in (2.1).

[21] "An Epic Debunking of the Saturated Fat Myth," *Authoritynutrition.com*, July 2015. https://authoritynutrition.com/it-aint-the-fat-people/ (Accessed 15 May 2015).

[22] "Fiber: Start Roughing It!," *Nutrition Source*, Harvard School of Public Health, n.d. https://www.hsph.harvard.edu/nutritionsource/carbohydrates/fiber/ (Accessed 15 May 2015). (Free to read.)

[23] "Cholesterol: Top 5 Foods to Lower Your Numbers," *Diseases and Conditions: High Cholesterol*, Mayo Foundation for Medical Education and Research, n.d. http://www.mayoclinic.org/diseases-conditions/high-blood-cholesterol/in-depth/cholesterol/art-20045192 (Accessed 15 May 2015). (Free to read.)

[24] "Black Beans," *World's Healthiest Foods*, George Mateljan Foundation, n.d. http://www.whfoods.com/genpage.php?tname=foodspice&dbid=2 (Accessed 15 May 2015). (Free to read.)

[25] Brill, J. *Cholesterol Down: Ten Simple Steps to Lower Your Cholesterol in Four Weeks—Without Prescription Drugs.* New York: Three Rivers Press (2006).

[26] Turner, N. D., and Lupton, J. R., "Dietary Fiber," *Advances in Nutrition*, 2, no. 2 (2011), 151–152. (Free to read.)

[27] "Feed Yourself Fuller," British Nutrition Foundation, 2010. https://www.nutrition.org.uk/attachments/423_13209%20BNF%20feed%20Poster_PRINT_2.pdf (Accessed 15 May 2015). (Free to read.)

[28] Ashwell, M., and Hsieh, S. D., "Six Reasons Why the Waist-to-Height Ratio Is a Rapid and Effective Global Indicator for Health Risks of Obesity and How Its Use Could Simplify the International Public Health Message on Obesity," *International Journal of Food Sciences and Nutrition*, 56 (2005), 303–307.

[29] Browning, L. M. et al., "A Systematic Review of Waist-to-Height Ratio as a Screening Tool for the Prediction of Cardiovascular Disease and Diabetes: 0.5 could be a suitable global boundary value," *Nutrition Research Reviews*, 23, no. 2 (2010), 247–269.

[30] Ashwell, M., et al., "Waist-to-Height Ratio Is More Predictive of Years of Life Lost than Body Mass Index," *PLoS One*, 9, no. 9 (2014). (Free to read.)

[31] Storrs, C. "Fat Is Back: New Guidelines Give Vilified Nutrient a Reprieve," *CNN*, 23 June 2015. http://www.cnn.com/2015/06/23/health/fat-is-back/ (Accessed 23 June 2015). (Free to read.)

[32] Tinker, B. "Cholesterol in Food Not a Concern, New Report Says," *CNN*, 19 February 2015. http://www.cnn.com/2015/02/19/health/dietary-guidelines/ (Accessed 23 Jun. 2015). (Free to read.)

[33] The quote is from "Geometry and Experience," an expanded form of an address by Albert Einstein to the Prussian Academy of Sciences in Berlin (27 January 1921). In "Albert Einstein," translated by G. B. Jeffery and W. Perrett, *Sidelights on Relativity* (1923).

[34] "Who Pays Taxes in America in 2015?," *Citizens for Tax Justice*, 9 April 2015. http://ctj.org/ctjreports/2015/04/who_pays_taxes_in_america_in_2015.php (Accessed 15 May 2015). (Free to read.)

[35] "Who Pays? A 50-State Report by the Institute on Taxation and Economic Policy," Institue for Taxation and Economic Policy, n.d. http://www.itep.org/whopays/ (Accessed 15 May 2015). (Free to read.)

[36] Cosgrove-Mather, B. "50 Years for the Golden Arches," *CBS News*, 15 April 2005. http://www.cbsnews.com/news/50-years-for-the-golden-arches/ (Accessed 15 May 2015). (Free to read.)

[37] "What Is Inflation and How Does the Federal Reserve Evaluate Changes in the Rate of Inflation?" Board of Governors of the Federal Reserve System,

26 January 2015. http://www.federalreserve.gov/faqs/economy_14419.htm (Accessed 15 May 2015). (Free to read.)

[38] "Why Does the Federal Reserve Aim for 2 Percent Inflation over Time?," Board of Governors of the Federal Reserve System, 26 January 2015. http://www.federalreserve.gov/faqs/economy_14400.htm (Accessed 15 May 2015). (Free to read.)

[39] U.S. Census Bureau, Census 2000 Brief, *Housing Costs of Renters: 2000* (C2KBR-21); 2010 American Community Survey (B25064). https://www.census.gov/hhes/www/housing/census/historic/grossrents.html (Accessed 15 May 2015). (Free to read.)

[40] Bostock, M., et al., "Is It Better to Rent or Buy?," *Upshot, New York Times*, n.d. http://www.nytimes.com/interactive/2014/upshot/buy-rent-calculator.html (Accessed 15 May 2015). (Free to read.)

[41] Bernard, T. S. "New Math for Retirees and the 4% Withdrawal Rule," *Your Money, New York Times*, 8 May 2015. http://www.nytimes.com/2015/05/09/your-money/some-new-math-for-the-4-percent-retirement-rule.html (Accessed 15 May 2015). (Free to read.)

[42] "The Discount Rate," Board of Governors of the Federal Reserve System, 26 May 2015. http://www.federalreserve.gov/monetarypolicy/discountrate.htm (Accessed 15 May 2015). (Free to read.)

[43] "Weekly National Rates and Rate Caps," Federal Deposit Insurance Corporation, 15 June 2015. https://www.fdic.gov/regulations/resources/rates/ (Accessed 15 May 2015). (Free to read.)

[44] "Consumer Price Index Frequently Asked Questions," Bureau of Labor Statistics, U.S. Department of Labor, 7 September 2014. http://www.bls.gov/cpi/cpifaq.htm#Question_1 (Accessed 15 May 2015). (Free to read.)

[45] Berkshire Hathaway. 2011 Annual Letter to Shareholders. http://www.berkshirehathaway.com/letters/2011ltr.pdf (Accessed 15 May 2015). (Free to read.)

[46] Jaconetti, C. M., et al., "Best Practices for Portfolio Rebalancing," Vanguard research, July. 2010. http://www.vanguard.com/pdf/icrpr.pdf (Free to read.)

[47] Davis, J., and Piquet, D., "Recessions and Balanced Portfolio Returns," Vanguard research, October 2011. (Free to read.)

[48] Drake, F., and Sobel, D., *Is Anyone Out There? The Scientific Search for Extraterrestrial Intelligence*. New York: Delta (1994).

[49] McGinty, J. C. "To Find Love Match, Try Love Math (Results Will Vary)," *Numbers, Wall Street Journal*, 13 February 2015. http://www.wsj.com/articles/to-find-love-match-this-valentines-day-try-love-math-1423842975 (Accessed 15 May 2015). (Free to read.)

[50] "Secretary Problem," Wikipedia, n.d. https://en.wikipedia.org/wiki/Secretary_problem (Accessed 15 May 2015). (Free to read.)

[51] Bruss, F. T. "A Unified Approach to a Class of Best Choice Problems with an Unknown Number of Options," *Annals of Probability*, 12, no. 3 (1984), 882–891. (Free to read.)

[52] See Theorems 1.2.2 and 1.2.3 in Gusfield, D., and Irving, R. W. *The Stable Marriage Problem: Structure and Algorithms*. Cambridge, MA: MIT Press (1989).

[53] Gale, D., and Shapley, L. S., "College Admissions and the Stability of Marriage," *American Mathematical Monthly*, 69 (1962), 9–15.

[54] Iwama, K., and Miyazaki, S., "A Survey of the Stable Marriage Problem and Its Variants," *Proceedings of International Conference on Informatics Education and Research for Knowledge-Circulating Society 2008*, 131–136. IEEE Computer Society (2008).

[55] Rinaldi, S., and Gragnani, A., "Love Dynamics between Secure Individuals: A Modeling Approach," *Nonlinear Dynamics, Psychology, and Life Sciences*, 2 (1998), 283–301.

[56] "Bargaining Problem: Nash Bargaining Solution," Wikipedia, n.d. https://en.wikipedia.org/wiki/Bargaining_problem#Nash_bargaining_solution (Accessed 15 May 2015). (Free to read.)

[57] Nash, J. F. Jr., "The Bargaining Problem," *Econometrica*, 18, no. 2 (1950), 155–162.

[58] Gottman, J., et al., "The Mathematics of Marital Conflict: Dynamic Mathematical Nonlinear Modeling of Newlywed Marital Interaction," *Journal of Family Psychology*, 13 (1999), 1–17.

[59] Gottman, J., et. al., *The Mathematics of Marriage: Dynamic Nonlinear Models*. Cambridge, MA: MIT Press (2002).

[60] Heyman, R. E., and Smith Slep, A. M., "The Hazards of Predicting Divorce without Crossvalidation," *Journal of Marriage and Family*, 63, no. 2 (2001), 473–479.

[61] Radboud University Nijmegen, "Mathematical Counseling for All Who Wonder Why Their Relationship Is Like a Sinus Wave." ScienceDaily, 15 November 2012. https://www.sciencedaily.com/releases/2012/11/121115132855.htm (Free to read.)

Index

■ ■ ■ ■ ■ ■ ■ ■ ■ ■ ■ ■ ■ ■ ■ ■ ■